THE TOTALIZING ACT:
KEY TO HUSSERL'S EARLY PHILOSOPHY

PHAENOMENOLOGICA
COLLECTION FONDÉE PAR H. L. VAN BREDA ET PUBLIÉE
SOUS LE PATRONAGE DES CENTRES D'ARCHIVES-HUSSERL

112

JONATHAN KEARNS COOPER-WIELE

THE TOTALIZING ACT:
KEY TO HUSSERL'S EARLY PHILOSOPHY

Comité de rédaction de la collection:
Président: S. IJsseling (Leuven)
Membres: L. Landgrebe (Köln), W. Marx (Freiburg i. Br.),
J. N. Mohanty (Philadelphia), P. Ricoeur (Paris), E. Ströker (Köln),
J. Taminiaux (Louvain-la-Neuve), Secrétaire: J. Taminiaux

THE TOTALIZING ACT: KEY TO HUSSERL'S EARLY PHILOSOPHY

JONATHAN KEARNS COOPER-WIELE
University of Massachusetts at Boston, U.S.A.

KLUWER ACADEMIC PUBLISHERS
DORDRECHT / BOSTON / LONDON

Library of Congress Cataloging-in-Publication Data

```
Cooper-Wiele, Jonathan Kearns, 1954-
   The totalizing act : key to Husserl's early philosophy / by
 Jonathan Kearns Cooper-Wiele.
       p.    cm.
    Bibliography: p.
    Includes index.
    ISBN 0-7923-0077-7 (U.S.)
    1. Husserl, Edmund, 1859-1938.  2. Whole and parts (Philosophy)-
 -History--20th century.  3. Act (Philosophy)--History--20th century.
 I. Title.
 B3279.H94C66   1989
 193--dc19                                                  88-38142
```

ISBN 0-7923-0077-7

Published by Kluwer Academic Publishers,
P.O. Box 17, 3300 AA Dordrecht, The Netherlands

Kluwer Academic Publishers incorporates
the publishing programmes of
Martinus Nijhoff, Dr W. Junk, D. Reidel and MTP Press

Sold and distributed in the U.S.A. and Canada
by Kluwer Academic Publishers,
101 Philip Drive, Norwell, MA 02061, U.S.A.

In all other countries sold and distributed
by Kluwer Academic Publishers Group,
P.O. Box 322, 3300 AH Dordrecht, The Netherlands

All rights reserved
© 1989 by Kluwer Academic Publishers
No part of the material protected by this copyright notice may be reproduced or utilized in any form or by any means, electronic or mechanical, including photocopying, recording or by any information storage and retrieval system, without written permission from the copyright owner.

Printed in the Netherlands

For my parents,
Laura Elizabeth Scott Wiele
and
Lester Henry Samuel Wiele
ἐν ἀρχῇ ἦν ὁ λόγος
"In the beginning was the word . . ."

Table of Contents

ACKNOWLEDGEMENTS	ix
INTRODUCTION: THE ORIGINS OF HUSSERL'S TOTALIZING ACT	1

Chapter
I. THE TOTALIZING ACT: KEY TO HUSSERL'S EARLY
    PHILOSOPHY   18
        The Totalizing Act   18
        The Totalizing Act as Totality   22

II. THE CONCEPT OF THE TOTALIZING ACT AS COLLECTIVE
    CONNECTION: PROGENITOR OF NUMBER   31
        The Auto-Abstraction of the Concept of Collective
            Connection   31
        Number Concepts: Progeny of the Totalizing Act   34
        The "Attachment" of Number Concepts: Index of the
            Totalizing Act   36
        The Preeminence of the Totalizing Act: Refutation of a
            Prevalent Interpretation   40

III. SYMBOLIZING: PROSTHESIS OF THE TOTALIZING ACT   47
        The Hierarchic Complication of Totalizing Acts   48
        The Anatomy of *Abstracta*   50
        The Self-Extension of the Totalizing Act by Proxy   54

IV. THE SYMBOLIC TOTALIZATION OF SENSIBLE MULTITUDES   61
        The Sensible Individual as Modified Multitude   62
        The Symbolic Totalization of the Sensible Multitude   63

V. THE INTUITIVE TOTALIZATION OF THE INDIVIDUAL SENSE
    OBJECT   72
        The Sensible Group: Sufficient Context for Analyzing
            Intuition of Individuals   74

viii   *Table of Contents*

       The Problem: Non-Convertibility of Simultaneous and Successive Totalizing   77
       The Resolution: Successive and Simultaneous Totalizing as Continuous   80
       The Mutual Implication of Intuiting and Representing in the Intuition of the Sensible Thing   83

VI. THE TOTALIZING ACT AS MEDIATOR OF THE IDEAL AND REAL   90
       Hypothesis: The Internal Motivation for the Great Inversion   91
       Confirmation: The *Prolegomena* of 1900   95

VII. THE ENSOULMENT OF SENSATION: TRIUMPH OF THE TOTALIZING PSYCHE   109
       The Immanent Object as Empiricistic Fetish   111
       The Psychical Production of the Transcendent Object   115
       The Dilemma: The Uncertainty of the Transcendent and the Imperceptibility of the Immanent   118
       The Great Reversal: The Causal World as Interpretation   121

AFTERWORD: A HYPOTHETICAL ANSWER FOR ALFRED SCHUTZ   126

APPENDICES   129

SELECTED BIBLIOGRAPHY   145

INDEX   149

Acknowledgements

Acknowledgement that one's thought and writing are virtually a confluence of many influences is incumbent upon the author who would be honest. Like Jastrow's "duck-rabbit" seen in one way, a book indicates the labors of its writer; seen in another and equally legitimate manner, it signifies a multitude to whom its author is indebted.

More than to any other Husserl scholar, I am indebted to Professor Theodore de Boer. My dialogue with his *The Development of Husserl's Thought* has never ceased to be rewarding.

I am grateful to my teachers at Boston University — Professors H. H. Oliver, W. Marx Wartofsky, J. N. Findlay, Erazim Kohák, and Bernard Elevitch — in whose courses my thought and expression were nurtured, cultivated, and weeded. Thanks are due to Professors Kohák and Oliver for recommending my dissertation to Kluwer Academic Publishers and to Professor Oliver for remarkable generosity and continued collaboration. Special thanks to Professor Robert Cohen for his kind interest in my work and for his advocacy and delivery of my manuscript to Kluwer Academic Publishers. The enthusiasm of Professor Judson Webb for this project at its inception, and his willingness to share his research concerning Husserl and Hilbert, was a valued inspiration at a critical time.

I have much appreciated the kindness and good will of Professor S. IJsseling of the Husserl-Archives, Leuven, in this my first experience with the publishing process.

Thanks are also due my friend, Dr. Gary Zabel, Socratic analyst and interlocutor *par excellence*; I am indebted to him for many aporetic conversations on Husserl, and for support through the most trying of times. I am grateful as well to Mr. Douglas Anderson, lover of Sartre, for generous friendship, and to Mr. Lance Carden, to whom spirited argument only draws me closer. Many thanks to Connie and Warren Ogilvie for providing the opportunity for a very happy and stable home in which my dissertation might be written. I am inexpressibly grateful to Laura E.

Wiele, my mother, who has known and supported me longer than anyone and whose faith in me has often restored my faith in myself.

I am deeply indebted in innumerable ways to my wife and companion, Beverly Y. Cooper-Wiele, who has been more than meet help. She has done all of the things well for which wives are typically credited in their husband's books: deciphered, typed, edited, proof-read; all tirelessly and with a humor and perspective I often lacked. More important, her relentlessly critical intellect is both inspiration and ideal; she wields naturally what I do only with rather more effort.

Finally, I owe a large debt of gratitude to the late Professor Dr. Edmund G. Husserl. My engagement of his work has been unfailingly provoking, illuminating, and inspiring. It was a very happy coincidence that I defended my dissertation, rooted in his *Habilitationsschrift* of 1887, in the centenary year of that work. It is my sincerest hope that he would accept this essay as some small repayment of my intellectual debt to him.

Jamaica Plain, Massachusetts   Jonathan Kearns Cooper-Wiele
August 9, 1988

INTRODUCTION

The Origins of Husserl's Totalizing Act

At noon on Monday, October 24th, 1887, Dr. Edmund G. Husserl defended the dissertation that would qualify him as a university lecturer at Halle. Entitled "On the Concept of Number," it was written under Carl Stumpf who, like Husserl, had been a student of Franz Brentano. In this, his first published philosophical work, Husserl sought to secure the foundations of mathematics by deriving its most fundamental concepts from psychical acts.[1]

In the same year, Heinrich Hertz published an article entitled, "Concerning an Influence of Ultraviolet Light on the Electrical Discharge." The article detailed his discovery of a new "relation between two entirely different forces," those of light and electricity.

Hermann von Helmholtz, whose theory guided Hertz's initial research, called it the "most important physical discovery of the century," and Hertz became an immediate sensation. He lectured on his discovery in 1889 before a general session of the German Association meeting in Heidelberg. In this lecture that, as he wrote beforehand to Emil Cohn, he was determined should not be "entirely unintelligible to the laity," Hertz explained that light ether and electro-magnetic forces were interdependent. He went on to tell his audience that they need not expect their senses to grant them access to these phenomena. Indeed, he said, the latter are not only insusceptible of sense perception, but are *false* from the standpoint of the senses.

The light and electricity of which he spoke, Hertz informed his listeners, are accessible only through the formulas of mathematical physics which reveal the unity underlying the flux of perceptual experience. These formulas, he continued, seem to "have an independent life and intelligence, as if they were wiser than we, even wiser than their discoverer, as if they give us more than he put into them."

Hertz gives voice in this lecture to what is an archetypal assumption in western thought from Empedocles to Galileo and Descartes, to Helmholtz himself. It provided the framework of presuppositions which gave rise to the method of psychological analysis employed by Dr. Edmund G. Husserl in his inaugural dissertation of 1887.

The assumption is that the world of ordinary sense perception is explained or accounted for by that which is itself imperceptible. This explanation must then be false, as Hertz asserted, if the senses are the criteria of truth and falsehood. And from the perspective of this explanation, the world perceived by the senses must be only apparent.[2] Some of Hertz's contemporaries, for instance, sought to explain his discovery in terms of current molecular theory. Yet they by no means supposed that the "sharply distinguishable forms of bodies in the phenomenal world" were correspondent to anything in the molecular world. Bodies having the appearance of solidity might "contain Graetz's 'fluid' molecules" or perhaps "Lehmann's 'liquid' crystals."[3]

Many such as Helmholtz were willing to embrace the objects of mathematical physics as "true reality," and draw the venerable conclusion that the world given to sense perception was to be epistemically dismissed as pure deception.[4] Others such as Ewald Hering, a former colleague of Carl Stumpf's at the University of Prague, opposed Helmholtz's conclusion.

Hering argued that many facets of the perceived world, colors for instance, were perfectly legitimate objects for description "regardless of their causal conditions." Hering's careful "analysis and systematic arrangement" of color phenomena led to such results as the "doctrine of the four basic achromatic colors, the 'natural system of colors,' the discovery of lustre and voluminousness" and the "distinction between color and brightness."[5] Those like Hering who advocated a more sober assay of the phenomena rendered by the senses did not necessarily reject the explanatory concepts and laws proffered to account for them. They often believed that such description was not merely desirable but a necessary propaedutic to explanation.

Others, like Johann Wolfgang von Goethe earlier, in his *Theory of Colors* of 1810, were more explicit in their rejection of the "Newtonian" imposition of mathematical constructs on the perceived world, an imposition that was often but a substitution. Goethe demonstrated that there was "more to perspective than geometrical models" by pointing out, e.g., that red objects obtrude and blue objects recede on a canvas. He was not unaware of the debates concerning whether light consisted of particles or waves but chose to extoll and investigate the perceptual given as more than could be explained. However, by the time Goethe's book was translated from the German in 1840, the theory of transverse waves of Fresnel and Arago had superceded corpuscular theory, and the "idea of abandoning a mathematical approach to optics would have seemed ludicrous."[6]

Nevertheless, there were others who neither accepted the "reality" accessed by the equations of which Hertz spoke as true and rejected the sensible world as a deception, nor did the opposite, nor conceded the validity of explanatory constructs but elaborated a descriptive method as a propadeutic to their formulation. Those such as Gustav Kirchhoff argued that physical concepts, e.g., those of "mass," and "force," and "atom," were valid as expedients in the higher level description of phenomena but not themselves actual "objects" displacing the world of ordinary sense experience.

Ernst Mach claimed that the concept of the atom could be "eliminated by describing the causal relations which emanated from" it in terms of "functional relationships expressed in differential equations." In his *Analysis of Sensations* of 1885 he attempted to account for all physical concepts in terms of sensations. Yet Mach relied heavily on physiological theory, accounting for sensations in terms of events in the central nervous system and brain. Even within the province of perceptions proper he maintained, e.g., that the immediate perception of the color orange actually involved a complex of sensations in the retina representing red and yellow.[7]

The world given to the senses and the putative physical world explanatory of it had long been sundered in western thought. Whether or not one believed in the latter to the detriment of the former, or vice versa, the sensible world continued to present itself in a more or less consistent fashion. If anything, the debates concerning the status and function of physical concepts, and the theories employing them, strengthened the conviction that careful description of phenomena accessible to everyone ought to precede explanatory efforts by mathematical physics. If the nature of "true reality" was uncertain, a more certain descriptive account of the phenomenal point of departure of its theories seemed both viable and desirable. Indeed, a clear grasp of the apparent could only facilitate an explanatory account of it.

This was the view of Franz Brentano, the teacher of Carl Stumpf, Thomas Masaryk, Alexius Meinong, Kasimir Twardowski, Anton Marty, Sigmund Freud, Edmund Husserl and others. Brentano initially took great hope in the emerging sciences and their method and believed they offered a way for philosophy beyond the collapse of speculative metaphysics. Brentano assumed the division of descriptive and explanatory labors and appealed to the sciences themselves for justification. There was, after all, a descriptive anatomy, geology (geonosy), and related petrography on the one hand, and an explanatory physiology, geology (geonomy), and related petrogenesis on the other.[8]

Brentano also assumed the perceptual world was but the appearance of an actually existent world underlying and giving rise to it. But he was also keenly cognizant of the debates surrounding the status of physical concepts, and he therefore refused to be much more than agnostic concerning the specifics of this imperceptible causal reality. He regarded sense experiences as signs announcing the presence of something producing them but conceded that the latter remained incognito. We can say that "*something is at hand,*" he asserted, when there is a sensational experience, "but this is as far as we can go. In and of itself, that which truly exists is not in the appearance, and that which appears does not truly exist." Sensible appearances "exist," then, "only phenomenally and intentionally."[9] Hertz asserted that reality accessible only through Maxwell's mathematical formulas was false from the standpoint of the senses. Brentano held that "outer perception" of sensible objects, produced in some sense by the "objects" of those formulas, is not worthy of the term "perception" or of "taking as true" (*Wahr-nehmung*). On the contrary, if one credits sensible appearances with actual existence, one has failed to see that they must be "taken as false" (*Falsch-nehmung*).[10]

Brentano sought certain and solid ground for philosophy in the wake of what he regarded as an age of decline after Bacon, Descartes, Locke, and Leibniz, culminating in the debacle of speculative metaphysics. Given the presuppositions with which he began to philosophize, he rejected the possibilities of both the sensible world and that of theoretical physics as suitable loci of certainty. The former is false, and the concepts of the latter are incorrigibly hypothetical: "That which truly exists is not in the appearance, and that which appears does not truly exist." Further, even the most general and mathematical laws of theoretical physics were regarded by Brentano as no more than "general facts." As such, they could never be more than highly *probable*, they remained the "truths of fact" of which Leibniz had spoken. And this was true not only of, e.g., Boltzmann's statistical theory of gas molecules but of non-statistical laws as well.[11]

Brentano thus sought epistemic refuge in the last remaining place where certainty might be had, one prepared by Descartes. He turned to the sphere of immanent and immediate consciousness not out of a Cartesian desire to found his reading of "Nature's book," as Galileo termed it, that book written in the language of "triangles, circles, and squares."[12] Brentano was rather more concerned to find a firm epistemic foundation in order to resolve the problem of immortality and that of whether there was a "divine source of all being."[13]

It is not surprising that he then demarcated the categories of phenomena in terms of the "psychical" and the "physical," the crucial distinction being that the former was "intentional." Both categories of phenomena fell within the domain of immanent consciousness. Within this domain Brentano established these rigid distinctions between psychical acts and their objects.

The locus of certainty was that of "inner perception," that awareness Brentano maintained accompanied every psychical act. As an object of "outer perception," the perceived horse, e.g., cannot be taken to be as it appears. Brentano held that there is nevertheless in the awareness attending such a perceiving, the incorrigible *certainty that* a horse is being perceived. Because this awareness or inner perception is one with the act, there is no chance of its being delusive. This certainty of the act of perceiving, regardless of the credibility of the perceived, is *evident*. Brentano's epistemic position is not one occasioned by the illusions and misperceptions found in ordinary perception but by his fundamental belief in imperceptible, explanatory reality (and in his agnosticism concerning its specific character).

The certainty of the perceiving of a horse is incorrigible, and this excludes the possibility that one's perception is not that of a horse. Brentano founds the apodictic judgment expressed as the law of non-contradiction on the awareness of the impossibility of the opposite of an evident perception, or judgment, being the case simultaneously. The concepts presumed by this judgment are derived in inner experience, and the certainty of the law, in contrast to those of theoretical physics, is attained "on the basis of a single clear observation." This produces "absolute certainty" of the "universal validity of the law, and as absolute, this certainty cannot in any way be increased." Brentano referred also to this absolute or apodictic certainty as "mathematical certainty" and opposed it to that "physical certainty" characterizing inductive laws of even extremely high probability. Brentano observed of the latter that "we have no absolute certainty, but we believe nonetheless that we ought to prefer the acceptance of the law to its rejection." As grounds for this preference he proffered the Humean rationale that "this belief is strengthened by the repetition of the same observation in many other cases."

Stumpf also pursued the charting of necessary structures and relations of the sensible givens of color and tones. In the tradition of his colleague Hering, he enunciated such laws of phenomena as that color must be extended, but the extended need not be colored.[14]

In the same way that inner perception was prescribed as the locus of certainty by presuppositions concerning the sensible and insensible worlds, Brentano and his students presumed the empirical structure of mind as a complication of simples and the analytical method appropriate to it. Under the influence of the new chemistry, the complications of consciousness were not, however, regarded as consisting merely of atomistic representations. Stumpf maintained with his colleague Lotze that it is "conceivable that out of two representations, a third and new should arise, which in no wise is the mere sum of the earlier ones; just as a chemical compound does not possess the sum properties of its elements, but rather properties that are new." Stumpf thus termed his endeavor to analyze psychical representations, including those abstract or conceptual, "psychological analysis." It was patterned after that of the chemist who seeks "to reduce composite matters, with which we commonly deal, back to their elements." Psychological analysis was not complete, in Stumpf's view, until concepts had been "clarified" by analyzing them into their constituent concepts. Utterly simple concepts could then be clarified only by indicating those representations from which they were abstracted.[15]

It was Brentano's descriptive psychological method, and the entire configuration of presuppositions surrounding and giving rise to it, that Dr. Edmund G. Husserl took up in his inaugural dissertation under Carl Stumpf at Halle. Brentano's philosophy must have seemed to Husserl the perfect fulfillment of predispositions acquired through his mathematical studies under Karl Weierstrass and Leopold Kronecker.

Husserl began his mathematical studies under these men in Berlin in 1878 at the age of 19, having studied in Leipzig since 1876. Under their tutelage, and especially that of Weierstrass, Husserl forsook astronomy and devoted himself entirely to mathematics and the philosoply of mathematics. He left Berlin for Vienna in 1881, where he completed his doctorate under Leo Königsberger but returned to serve as an assistant to Weierstrass upon its completion in 1883.

Husserl later acknowledged the "lasting influence" of Weierstrass on him and stated, "I got the ethos of my scientific striving from Weierstrass." It was the latter who aroused in the young Husserl "an interest in a radical grounding of mathematics." Weierstrass's goal was no less than the rationalization of mathematical analysis. His method, according to Husserl, was to determine its "original roots, its elementary concepts and axioms." Having laid these foundations, the whole system of analysis could then "be constructed and deduced by a fully rigorous, thoroughly evident method."[16]

The elementary concepts of which Husserl spoke were those of the positive integers of elementary arithmetic. Weierstrass's "arithmeticization" of analysis grew out of his deep distrust of conclusions drawn on the basis of "physical experience." He demonstrated that certain propositions deemed correct on the basis of "geometrical intuition" — e.g., that "a continuous curve necessarily possesses a tangent, except perhaps at certain isolated points" — were in fact incorrect. He consequently rejected geometry as a suitable basis for mathematical analysis and pursued investigations toward this end in arithmetic instead.[17]

Weierstrass penetrated to the very origins of arithmetical concepts in these investigations. It is clear from Husserl's notes on his teacher's lectures in 1878 and 1880 that Weierstrass derived the "concept of number" from the operation "which we call enumerating" (*Zahlen*).[18] This operation is the gathering of a determinate "plurality of unities" from which is derived the concept of number. It is unclear from Husserl's notes whether or how Weierstrass distinguished the concept of the number from the particular plurality of entities.

Weierstrass fell ill in the winter semester of 1883—84, and Husserl left Berlin, served a year in the military, and then returned to Vienna to hear Brentano, both out of curiousity and at the urging of his friend and former pupil of Brentano, Thomas Masaryk. Husserl recounted that it was at this juncture that "I hesitated between remaining in mathematics as my life's vocation and devoting myself entirely to philosophy." The latter had been his minor area of study for the doctorate. His hesitation was short-lived, for he stated that "Brentano's lectures settled the issue." It was apparently Brentano's unswerving conviction that philosophy could itself be pursued with scientific rigor that persuaded Husserl to devote himself to it. He attended Brentano's lectures until the end of the winter semester of 1885—86 when, in the words of Carl Stumpf, "Husserl, recommended by Brentano, came (to Halle) . . . for the purpose of habilitation and became my student and friend."[19]

Husserl could hardly have found the psychological analysis practiced by Brentano and Stumpf anything but tailored precisely to his philosophical and mathematical predispositions acquired largely from Karl Weierstrass. As the title of his inaugural dissertation testifies, his interest in securing the foundations of mathematics in arithmetic had not abated. Indeed, it was very likely strengthened by his encounter with Brentano's philosophy.

Weierstrass's endeavor to arithmeticize analysis was motivated by his conviction that geometrical or idealized spatial intuitions could not be trusted. Brentano was persuaded that hypothetical concepts and spatio-temporal appearances of the imperceptible causal world were equally untrustworthy. Weierstrass's consequent recurrence to arithmetic, to the concept of number, and especially to the psychical act of counting as the source of this concept, paralleled Brentano's recourse to immanent consciousness and to the acts within it. Weierstrass's penetration to the most elementary concepts, and even to their sources in intuition, was certainly congenial to the similar method of psychological analysis advocated by Brentano and Stumpf. Husserl pointed out that Weierstrass's endeavor to rationalize mathematics was tantamount to its deduction from these arithmetical foundations "by a fully rigorous, thoroughly evident method." This mathematical evidence was, as Brentano recognized, equivalent to the apodictic certainty he sought. It was unequivocally different in kind from the certainty of high probability associated with inductively derived laws.

Husserl's dissertation of 1887 at Halle represents the confluence of the respective influences of Weierstrass and Brentano on his early thought. What is incipiently unique is due to the locus of this confluence, the emerging thinker in his own right, Edmund Husserl.

In continuing Weierstrass's project, Husserl placed the act of enumerating at the center of his dissertation. In doing so, he simultaneously embraced Brentano's conviction that the locus of apodictic certainty was the psychical act and its attendent inner perception. There was virtually no other place Husserl could have begun given the framework of assumptions inherited from his mentors. This framework posited an abyss between, on one side, the psyche and its acts, apodictic or mathematical certainty, and logical and mathematical norms, and on the other side, (false) spatio-temporal particulars, hypothetical physical concepts indicating hidden causal machinations producing them, and inductive laws characterized by probability.

It was within this framework of presuppositions that Husserl began to philosophize. What is unique in his philosophizing emerges out of his beginning with the act of counting as the source of the concept of number, and the immediate implication of Brentano's epistemic-ontic presuppositions as its context. Husserl's beginning with this act introduced a mathematic voluntarism and creativity into Brentano's already voluntaristic psychical acts. It seems that Brentano imported the quality of "striving" as characteristic of the "intentional" act from the connative sphere of Thomas Aquinas who maintained: "*Intentio est proprie actus voluntatis.*" Kline argues that after 1500 mathematics became increasingly "dominated by concepts derived from the recesses of human minds." These were much more "creations of the mind" than "immediate idealizations or abstractions from experience."[20] Husserl's own teacher Kronecker witnessed to this by exclaiming that "God made the integers, everything else is man-made." And Christoph Sigwart, in his *Logic* of 1873, linked the "voluntary thought presupposed in logic" to questions of concrete moral ends.[21] Husserl also spoke of numbers as "mental creations" in his inaugural dissertation.

Husserl's originality does not lie however in his mere assumption of this voluntaristic, intentional act. It is the thesis of this essay that what is unique in his early work, beginning with his dissertation of 1887, is that he not only accounts for Weierstrassian pluralities in terms of this activity, but progressively and inexorably for "pre-collected" sensible multitudes and their individual members as well. This inexorable progression is the thread uniting Husserl's early work and that which ultimately effects his "transcendental turn."

This is not to imply that Husserl's conceptual project of founding mathematics and logic is peripheral. It might be said that the founding of the concept of number exerts a certain "demand" on the act of counting until a critical juncture when the "conceptual" project diverges from that of the "sensible," and the latter takes on a life of its own. The two projects converge again in the *Logical Investigations* preceded by an analogous demand on the psychical act, not by concepts but by Ideas.

These "demands" are *not* sufficiently unilateral to justify the conclusion that the conceptual project is the most significant in Husserl's early work. These "projects" are of a piece in that both effect a conquest of spatio-temporal phenomena. On the one hand, Husserl secures from psychical acts non-inductively derived concepts constituting apodictically certain propositions.

On the other, he effects a progressive accounting for spatio-temporal phenomena in terms of the psychical act. The ultimate result is reduction of the sphere of spatio-temporal particulars, even that of the causal reality underlying them, to the status of *interpretations* of immanent sense material by psychical acts. This is accomplished through a collaboration of the psyche and ideal meanings, but it renders the project culminating in the turn to the latter somewhat superfluous. The insensible causal and phenomenal worlds are now but interpretations by consciousness.

Husserl's unique contribution, made entirely within the framework of Brentano's and Stumpf's assumptions concerning consciousness and method, is the dissolution of the threat posed by the assumption of causal reality, spatio-temporal particularity, and the probability of the laws characterizing both. What Brentano took as a real threat, necessitating his seeking epistemic refuge in immanent consciousness, Husserl ultimately reveals as threatening *only insofar* as it is forgotten to be but a production of consciousness itself. Spatio-temporal phenomena, and the causality governing them, are no less uncertain or hypothetical qua *interpretations* by consciousness. Consciousness, however, becomes no mere refuge from this uncertainty but the reason for it. Consciousness therefore becomes the locus of apodicticity *par excellence*. In his *Logical Investigations*, Husserl criticizes Brentano's empiricist epistemology unsparingly for failing to account for the sensible phenomenal object in terms of consciousness itself.

Husserl is clear, in his introduction to "On the Concept of Number," that the failure to found the edifice of mathematics logically, to persist in employing concepts not rigorously deduced, mocks the pretensions of mathematics to paradigmatic rationality. This lack of rational justification consequently pervades scientific practice. The cleansing of mathematical and logical language of all arbitrariness was undoubtedly a moral crusade in Husserl's view. The philosophical ideal of the community of reasoners remains unfulfilled until the canons of reason are themselves rationalized.

There is no reason to assume that Husserl regarded the effects of such rationalization as any less practical than did Brentano, or Meinong, or Mach. As he insisted at length in his *Prolegomena* of 1900, the criteria governing thinking must be apodictically, not inductively, founded. These must be criteria by which thinkers could be convicted not of improbability but of absurdity.[22]

The conceptual "demand" upon the totalizing act is that it be the source of concrete collections if it is to be the source of the concept of number.[23] The concept of number, which Husserl believed to be the foundation of mathematics at this point, must derive from the psychical act. The latter, set over against spatio-temporal phenomena, hypothetical physical concepts, and the probability of induction, was the locus of apodictic certainty. As psychical, it was entirely external to and untainted by "physical" phenomena.

Husserl demonstrates in his dissertation, and in its subsequent form as roughly the first four chapters of his *Philosophy of Arithmetic* of 1891, that the "totality" of entities, sensible or not, is *tantamount* to their embrace by the totalizing act. It was essential that Husserl demonstrate that the totality *is* the totalizing act if this act was to be the abstractive base for the concept of number. The totality is *not* an object or content of the act. All that remains over against the latter are the entities totalized, and these are together only due to their selection by the act.

Husserl's demonstration of this point has not been evident to most interpreters. It is therefore necessary that his demonstration of it be itself demonstrated in Chapters I and II of this essay (a more detailed engagement of other interpretations is found in the footnotes and in the appendices they cite). Husserl's doctrine of the "attachment" or application of number concepts to concrete groups, after their production by psychical acts, is elaborated as further evidence of their derivation from acts alone. Number concepts are, in Husserl's view, as external to sensible groups as psychical acts. The significance of this doctrine of attachment has also been largely overlooked.

While the number concept, seeking derivation from the totalizing act, "demands" that this act be the sole source of totalities, it is more correct to say that the concept of number *may* be abstracted from this act, in Husserl's view, *because of the nature of the act itself*. It is because this act *is* the totalizing unity of entities it collects that the concept of number may ultimately be derived from it. The nature of this act as spontaneous and arbitrary, formalizing and formal, unified, as undetermined and therefore governed only by its own immanent teleology, will be elaborated in Chapters I and II. The centrality of the totalizing act to Husserl's *conceptual* project is decisively confirmed by the fact that number concepts are concepts *of* this act. Not only is it their abstractive base, but psychical acts are progenitors of these concepts at all points.

The totalizing act is quite limited in its capacity to hold each member of a totality in view along with all the others simultaneously. The concepts of such "actual" graspings are the actual number concepts. Husserl seeks to account for all of arithmetic, most of which is "inactual" or symbolic, in terms of the totalizing act. It is argued in Chapter III, concerned with portions of the *Philosophy of Arithmetic* other than those paralleling the inaugural dissertation, that Husserl also treats symbolic number concepts as concepts *of* totalizing acts. These acts are *actually* impossible, the totalizing psyche is finite. Yet its totalizing power is manifest in its ability to press proxies into service, signs or symbols by which it "constructs" the concepts of acts it cannot perform. Symbolizing activity functions, it is argued, as the "prosthesis" of the totalizing intent. This symbolic strategy is adumbrated in the inaugural dissertation when Husserl asserts that all contents of which there is an awareness are either actually in consciousness, or represented by those that are.

Husserl's conceptual project is virtually complete in the *Philosophy of Arithmetic*. The anatomy of its concepts is described in Chapter III, which prepares the way for Chapter VI and its treatment of Husserl's Ideatic turn. It is argued in the latter chapter that Husserl, having abstracted concepts from psychical acts, must have seen that these concepts referred not at all to psychical acts, and that the procedure of abstraction could not be recapitulated in reverse. It is maintained that this problem in the conceptual theory found its solution — the Ideatic turn — *in the very nature of the "concepts"* produced by psychical acts.

The doctrine of "propositions in themselves" was extremely popular both on the continent and across the channel, and one need not look far for possible influences on Husserl to account for his embracing of Ideas. This question of influence is not pursued for two reasons. First, it is the thesis of this essay that Husserl's early philosophy (as well as his transcendental turn) is best understood in terms of the centrality of the totalizing act. And second, it is the conviction of this writer, demonstrated in Chapters III and VI, that there is sufficient *internal* motivation in Husserl's own thought for his Ideatic turn. This is not to imply Husserl thought in a vacuum or to gainsay the affect on him of those like Frege, Bolzano, and Lotze. However, Dagfinn Føllesdal correctly observes, having himself argued for the decisive influence of Frege, that "even where there are similarities and contact, there is not always influence."[24]

The crucial juncture in Husserl's early thought occurs in the *Philosophy of Arithmetic* when he asserts that the primordial intent of the totalizing act is to totalize pre-given sensible multitudes when encountered by them.

Since the multitudes Husserl describes cannot be grasped actually, this intent may be the counterpart to the symbolic arithmetical accounting for inactual number concepts. It is less significant that Husserl invokes the gestalt quality of the multitude as the means by which the psyche symbolically grasps it, than it is that this grasp signifies the means by which the psyche totalizes that which it cannot totalize actually.

Even more significant, according to Chapter IV, is that Husserl describes the sensible *individual* within the context of the multitude. While his intent is to distinguish the two, what is most striking is the similarity of his approach to both. Husserl thinks of the sensible individual as more or less a multitude or collection of parts. There is precedent for this approach, for in Chapters I and II it is pointed out that he demonstrates that relations binding sensible individuals may only be articulated through a sort of reverse totalization. The totalizing act is witnessed also in his description of the way a number of similar characteristics are held in view prior to the abstraction of the concept uniting them. The totalizing act articulates the non-psychical or "physical" relation of similarity as well.

Chapter V proffers an interpretation of Husserl's article of 1894 contrary to that of most commentators including Husserl himself in a review three years later. In this article, it is argued, Husserl accounts for the immanent sensible object by employing the totalizing act in the same way it was used to account for the symbolic grasp of sensible multitudes. In doing so, he partitions immanent consciousness into that which is merely immanent *qua* noticed or sensed, and that which is *actually* immanent in virtue of having been intuited. The act of intuiting is structured in terms of the totalizing act. Husserl was willing to allow the sensible multitude to be accounted for symbolically, but he is not so willing in the case of the sensible individual. He is not for the simple reason that perception or intuition must not be symbolization. The object must be actually and immediately grasped. While Husserl adapts the totalizing act to fit this situation, he does so in such a way as to elucidate a latent dimension of the act rather than alter it.

It is argued further, in Chapter V, that symbolic or "representative" activity and totalizing activity virtually interpenetrate in Husserl's account of the intuitive immanentalizing of the sensible object. Intuition is impossible in this case without "*Repräsentation*." While Husserl's description of the way a sensible object becomes significative through an alteration in consciousness is of interest (what he and most interpreters have maintained is the article's central thesis), it is held that this insight is already more or less present in the *Philosophy of Arithmetic*. That Husserl did not acknowledge the rendering of the sensible individual actually immanent, via intuition, as the most important accomplishment of the article may be accounted for in more than one way. It may be that he was simply unaware of its significance and thus largely unaware of the progressive incursion of the totalizing act into sensibility. It is more likely that by the time of his review of the article in 1897, he had already become critical of this doctrine as he is in the *Logical Investigations*. If so, perhaps he was disinclined to acknowledge it as significant. But even if this is true, Husserl may have himself nevertheless been unaware of the integral role this analysis plays in his accounting for the *transcendence* of the perceived object in the *Logical Investigations*.

In Chapter VII it is suggested that Husserl's Ideatic turn, the logical conclusion of his conceptual project, occasioned a "demand" on the psychical act analogous to that discussed with reference to the inaugural dissertation. The totalizing act in the *Logical Investigations* is no longer the "form of the collection" but rather the means by which ideal numbers *qua* species are realized in its totalizing activity. The totalizing act mediates the ideal to the real. Since the individuation of the species *qua* form of the collection is contentual, this may have constituted a "demand" in Husserl's view that the individual members of the collection be transcendent rather than immanent. Or, perhaps, Husserl's rejection of the empirical theory of abstraction of concepts occasioned his attendant rejection of the empirical epistemology of the sensible object.

It is argued that Husserl overstates his rejection of the doctrine of the immanent object in his "Psychological Studies for an Elementary Logic" for two reasons. First, he accounts there for the immanent object *in terms of* consciousness and, specifically, of the totalizing act. Although in the *Investigations* he seeks to account for the object perceived as *transcendent*, this *remains* an accounting in terms of psychical activity. Second, and more important, the integral role of representation in intuition is carried over from the analysis in the earlier article to that of the *Investigations*. "*Repräsentation*" is fundamental in the latter because Husserl could not account for that not "in" consciousness in any other way than by that which was.

Chapter VII maintains that Husserl's adherence to this symbolic or representative doctrine in accounting for transcendent sensible objects requires his positing the meaningless sense data and the acts ensouling or animating them with ideal meaning. These "interpreted" sense data then appear to perceiving consciousness as the transcendent object. But this "object" is an *interpretation* only, one which cannot ultimately be determined correct or incorrect. Husserl consequently holds that there is certainty only of the sensations and interpretive acts which, *qua* immanent, may be perceived in "psychological reflection." This is at best dubious. Husserl is left with an interpretation of a transcendent object that cannot be verified and with immanent data that cannot be perceived. He is left at best with an interpretation that is seen in inner reflection. If so, Husserl fails to gain more than an immanent "object," a failure for which he castigated Brentano. Out of this circumstance grew his transcendental turn.

In rendering the sensible world an *interpretation*, Husserl's "conceptual" and "sensible" projects coincide once again. The interpretive act is a conferring of ideal meaning on meaningless sensations, and the representative accounting for transcendence evolved as a mode of the totalizing act. In the coincidence of Idea and act, causal reality itself becomes an interpretation of consciousness. It is argued that immanent sensational data are not therefore understood by Husserl as residue of external impingements on sense organs. The latter are *themselves interpretations of* sensations. According to Husserl, it is an illegitimate transgression of epistemological boundaries to assume that the interpreted *is* as it appears.

This is true *a fortiori* of "psychical" sensations such as pain. Husserl insists that this sensation is only *interpreted* as localized in some bodily member. If one believes that pain is actually coursing through one's arm, then one has exceeded the bounds of epistemic propriety in forgetting that this belief is founded only on an *interpretation upon an interpretation*. One's perceived arm is itself the result of an "objectifying interpretation" of immanent sense data by a psychical act, and the localizing of pain within this interpretation is also an objectifying interpretation of the sensation.

Husserl thus maintained Brentano's essential distinction between the "psychical" and "physical," between psychical acts and experiences and spatio-temporal particulars. Husserl's contribution is that of rendering these phenomena interpretations by immanent psychical activity and, by implication, the hypothetical-causal world an interpretation by such as well. The body of the individual consciousness is itself constituted in consciousness and the dependence of the "*psyche*" upon the "*soma*" is consequently reversed.

The bodies of others are no less interpretations in the individual consciousness; their "expressions" by means of them, e.g., of pain, are multi-layered tissues of interpretations within the individual consciousness.

Husserl regarded himself in the *Logical Investigations* as having broken through to the constitutive activity of consciousness only adumbrated by the British empiricists. His persistent criticism of them is not that they attempted to account for sensible experience in terms of consciousness but that they *failed* to do so. They simply imported the sensible object into the immanent sphere, assuming they had thereby accounted for it in terms of conscious activity.

It is the view of this writer that Husserl is correct *concerning the origins* of the position he reaches in the *Investigations*. And one cannot help but see him on the verge of his transcendent turn at this point. This turn occasions investigations of the constitution of the body and of the other consonant with those alluded to above. In the course of these investigations Husserl details data of "immediacy" that conflict with the theory of constitution. In the second volume of his *Ideen*, e.g., he insists that pain is "seen" immediately in the expressions of another, that touch reveals one's body and the surface touched as virtually co-emergent. Alfred Schutz was a follower of Husserl who asserted that he was unable to reconcile the constitutive theory with such data of concrete intersubjectivity, and that he would simply hold the former in abeyance while investigating the latter.[25]

One of the hypotheses emerging from this essay, set forth in the Afterword, is that the position Husserl reaches in the *Investigations*, as well as in the transcendental turn emerging out of them, is, as he acknowledges, the logical culmination of the essentially empiricistic presuppositions with which he began. He has simply radicalized them. As has been shown, Brentano's recurrence to consciousness was due essentially to his assumption of a causal reality productive of subjective sensational material. This was an assumption Husserl later rejected in, e.g., his *Krisis*, on the basis of an historical analysis of the origins and development of geometrical constructs. He argues in later works that Galileo but persisted in the mistaken hypostatization of these constructs in his construance of them as somehow behind and productive of sensible phenomena.

It is the conviction of this writer that Husserl's position in the *Logical Investigations* represents his carrying to its logical conclusion the epistemological framework which he later rejects. And he rejects it for other and better reasons than that phenomenal and causal reality is an interpretation in consciousness. If this is true, then perhaps there is both a resolution of Schutz's dilemma and a supercession of individual totalizing consciousness.

NOTES

1. Edmund Husserl, *Husserliana*, Band XII, *Philosophie der Arithmetik*, ed., Lothar Eley (The Hague: Martinus Nijhoff Publishers, 1970), p. 522. Theodore de Boer, *The Development of Husserl's Thought*, trans. Theodore Plantinga (The Hague: Martinus Nijhoff Publishers, Phaenomenologica, vol. 76, 1978), pp. 98f.
2. Christa Jungnickel and Russell McCormmach, *Intellectual Mastery of Nature, vol. 2, The Now Mighty Theoretical Physics, 1870–1925* (Chicago: The University of Chicago Press, 1986), pp. 85–98.
3. *Ibid.*, p. 101.

16   Introduction

4. David F. Lindenfeld, *The Transformation of Positivism: Alexius Meinong and European Thought, 1880—1920* (Berkeley: University of California Press, 1980), pp. 79f.
5. H. Spiegelberg, *The Phenomenological Movement*, vol. 1 (The Hague: Martinus Nijhoff Publishers, Phaenomenologica, vol. 5/6, 1971), p. 60.
6. David Knight, *The Age of Science* (Oxford: Basil Blackwell, Ltd., 1986), pp. 55, 64.
7. Lindenfeld, *Transformation of Positivism*, pp. 79f. The later nineteenth century witnessed a proliferation of specialities within physics which, in sub-dividing the discipline, left the impression of a less than unified scientific community, if not theory. There were many conflicting theories, e.g., in electrodynamics. Jungnickel and McCormmach, *Intellectual Mastery of Nature*, pp. 101, 110. Knight, *Age of Science*, pp. 4f.
8. Spiegelberg, *Phenomenological Movement*, p. 37.
9. de Boer, *Development of Husserl's Thought*, pp. 41, 18.
10. *Ibid.*, p. 36.
11. *Ibid.*, p. 78.
12. Aron Gurwitsch, *Phenomenology and the Theory of Science*, ed. Lester Embree (Evanston: Northwestern University Press, 1974), p. 45.
13. Spiegelberg, *Phenomenological Movement*, pp. 34, 29.
14. de Boer, *Development of Husserl's Thought*, pp. 36, 53, and 78f. Spiegelberg, *Phenomenological Movement*, pp. 35, 56, 63.
15. Dallas Willard, *Logic and the Objectivity of Knowledge* (Athens: Ohio University Press, 1984), p. 34. Lindenfeld, *Transformation of Positivism*, pp. 18f. See also Spiegelberg, *Phenomenological Movement*, vol. 1, pp. 42, 56, and 92. Brentano borrowed his category of "representations" from Descartes, and Stumpf, Husserl, and Meinong were versed in the British empiricists, even as first readings in philosophy.
16. Karl Schuhmann, *Husserl-Chronik: Denk-und Lebensweg Edmund Husserls* (The Hague: Martinus Nijhoff Publishers, Husserliana Dokumente, vol I, 1977), p. 7. See J. Philip Miller, *Numbers in Presence and Absence: A Study of Husserl's Philosophy of Mathematics* (The Hague: Martinus Nijhoff Publishers, Phaenomenologica, vol. 90, 1982), pp. 23—4. Also de Boer, *Development of Husserl's Thought*, pp. 97f.
17. Carl B. Boyer, *The History of the Calculus and its Conceptual Development* (New York: Dover Publications, Inc., 1949), pp. 284f.
18. Husserl, *Philosophie der Arithmetik*, p. 24, n. 1.
19. Dallas Willard, *Logic and Objectivity of Knowledge*, p. 32. The establishment of a psychology yielding propositions of which there was "absolute certainty" was no less than a moral and social mission for Brentano. He regarded his psychology as the discipline that would lay the foundations for future social progress and reformation. See de Boer, *Development of Husserl's Thought*, p. 104. His belief in evidential certainty accessible to all only contributed to his anti-authoritarian positions. As a pacifist, he opposed Prussian militarism and Bismarckian "Realpolitik." He was a Catholic priest, but was excommunicated upon the presentation of his brief to the German Bishops in Fulda attacking the proposed dogma of papal infallibility. It was his deep sense of virtually messianic mission that impressed students like Stumpf and Husserl, and they remembered him for it years afterward. Indeed, Stumpf even entered seminary due to his influence, and Anton Marty took Holy Orders. Spiegelberg, *Phenomenological Movement*, pp. 28f.

Meinong was deeply interested in educational and curriculum reform and collaborated with Höfler on the writing of textbooks. In this they were also joined by Mach, who understood his clarification of physical concepts as productive of social edification. He advocated the deconstruction of such concepts as "ego" and "force" in the education of children, believing this would initiate the end of social strife. He also allied himself with various Austrian and Russian socialists. Lindenfeld, *Transformation of Positivism*, pp. 79f.

Introduction 17

Brentano's descriptive psychology increasingly became for him less a propadeutic to explanation (as it functions in his *Psychology from an Empirical Standpoint* of 1874) and more the definitive means to elaborate evidential moral norms rather than those strictly logical and mathematical (as in his lecture of 1889 on the origins of ethical knowledge). It became the means to combat historicistic relativism infecting ethical theory and that of jurisprudence. See Spiegelberg, *Phenomenological Movement*, p. 38, and de Boer, *Development of Husserl's Thought*, p. 57.

20. Morris Kline, *Mathematics: The Loss of Certainty* (New York: Oxford University Press, 1982), pp. 167f. Kline expresses both Husserl's purpose and means when he observes that the mathematical work of the seventeenth through nineteenth centuries was, in being logically unfounded, "surely crude," but "also masterfully creative."
21. Willard, *Logic and the Objectivity of Knowledge*, p. 258.
22. In a manuscript dated between 1887 and 1890 (*KI* 28), Husserl cites the "essential difference" between the "objects" of mathematics and those of the natural sciences. Whereas mathematics deals only with "purely logical dependencies of magnitudes and positional relationships" which are "immediately given," the sciences deal with "real relationships of things to each other" which are only "hypothetical and inferred." Whereas every mathematical law is "subject to the principle of contradiction," and whoever denies it "is asserting an absurdity," a natural law is subject only to the "precepts of induction." One who denies the latter asserts that which is "in the optimal case, infinitely improbable," only. See Miller, *Numbers in Presence and Absence*, p. 8.
23. While to the knowledge of this writer Husserl does not designate this act as the "totalizing act," it will be so designated in the following chapters. The rationale for this is two-fold. First, it allows the correlation between it and the "*Inbegriff*," rendered in English as "totality," to be made explicit in English. This rendering conveys its original sense in Husserl's work of adding together. English employs such as "to total," to "tote up," as well as "sum total" and "total." Second, this rendering conveys some sense of the thesis of this essay that Husserl increasingly employs this act to account for sensibilia represented in consciousness. This "accounting-for" amounts to a domination of the psychical sphere by this totalizing act, the sense of which is perhaps conveyed better by the term "totalitarian." See Appendix I.
24. Hubert L. Dreyfus, ed., *Husserl, Intentionality, and Cognitive Science* (Cambridge: The MIT Press, 1984), p. 56. See preceding pages for Føllesdal's discussion of Frege and Bolzano in response to his on-going exchange with Mohanty on the question of influence. See Chapter VI below for a note on the possible resolution of the Føllesdal-Mohanty controversy. Cf., Willard, *Logic and the Objectivity of Knowledge*, pp. 178f.
25. Alfred Schutz, *Collected Papers, III: Studies in Phenomenological Psychology* (The Hague, Martinus Nijhoff Publishers, Phaenomenologica, vol. 23, 1970), pp. 51—52ff.

CHAPTER I

The Totalizing Act: Key To Husserl's Early Philosophy

THE TOTALIZING ACT

The totalizing psychical act, which is the key to Edmund Husserl's philosophy, first appears in his Inaugural Dissertation at Halle of 1887, titled "On the Concept of Number" (*Über den Begriff der Zahl*; hereafter *BZ*).

Husserl's objective in the dissertation was "the analysis of the concept of number."[1] As is evident in its introduction, this objective was set by another, which was the "clarification" (*Aufhellung*) of the "mysterious" and obscure "auxiliary concepts" (*Hilfsbegriffe*) of the imaginary, irrational, integral, differential, and continuous. These are, he laments, "apparently full of contradiction."[2]

Husserl's procedure for clarifying these concepts presumes a complicated hierarchy of mathematical concepts of which he assumes they are part. This conceptual hierarchy is composed of concepts which are "fundamental" and "underlying" and those which are "mediated." The former are described as "simpler" and "logically earlier," the latter as "more complicated" and "dependent." Those more complicated are consequently more "artificial" or "synthetic" (*künstlich*).[3]

The auxiliary concepts with which Husserl is most concerned are extremely complicated artificial concepts. He assumes that the higher analysis of which they are parts is itself derivative from elementary arithmetic. He further assumes that the latter has its "exclusive foundation" in the concept of number, in the "continuous series of concepts which mathematicians call the 'positive whole numbers.'"[4] The clarifying of these auxiliary concepts is then tantamount to their systematic analysis through the conceptual hierarchy, terminating only in fundamental and simple concepts. Husserl alludes to the "false views" of these concepts which entailed "many momentous errors" in mathematics. Their clarification will simultaneously clarify the practice and results of the sciences dependent upon mathematics.[5]

Husserl was not only concerned to clarify and analyze concepts constituted by other concepts. The fundamental concepts must also be clarified through reference to the "first and most simple manners of composition of psychical representations" (*Vorstellungen*).[6] The origin of these elementary concepts can only be clarified by pointing to the particular phenomena from which they are abstracted.

This conceptual hierarchy is paralleled by a hierarchy of psychical representations. The structure of this "highly intricate fabric of thought" is also ultimately clarified only by identifying the "simple representations" at its base. They are the "keys" to understanding the "higher levels of complication" with which thought "constantly operates, as with uniform and 'hardened' formations." These simplest constituents of psychical life are sensations.[7] Husserl employs the method of descriptive psychology as his "aid" (*Hilfsmittel*) in this analytical task.

Like his mentor, Karl Weierstrass, Husserl began his attempt at "arithmeticizing" mathematical analysis by assuming that the concept of number was to be derived from the psychical act of counting or colligating.[8] While he ultimately rejected both the view that analysis derived from arithmetic, and number from the psyche, the totalizing act continued to exert a decisive influence on his thought. It is the heart of *BZ*, and Husserl characterizes it in a number of ways mutually complementary and together circumscriptive of it.

First, the totalizing act is wholly "spontaneous. Its execution is entirely "discretionary" (*beliebig*), and it is therefore governed or motivated solely by "will (*Willkür*) and interest."[9] This spontaneous act is by no means anonymous, for Husserl identifies its subject as "we."

Second, since the subject "attaches" (*knüpfen*) these spontaneous activities to the contents, the totalizing act is *external* to them.[10] This externality of the act to its contents is one of its most significant characteristics. Unfortunately, its neglect by interpreters of *BZ* seems commensurate with its importance. Husserl clarifies the manner of this connection by describing the totalizing act as "concentrated" or "trained" (*richten*) upon the contents.[11]

Third, the latter characterization of the connection implies an "undivided attention" on the part of the act and its subject. This implication is made explicit, for Husserl states that totalizing activity is "concerted" or "undivided" (*einheitlich*).[12]

Fourth, the act's unity is expressed not only in its undivided attention to its contents, but also in the specific affect of its activity upon them. Its motivating "interest" virtually enacts itself in that Husserl describes the act as one of "unifying interest."[13] In virtue of this interest the particular contents of the act themselves become unified, "they are *one*."[14]

Fifth. There is no question that the act bears sole responsibility for the unity of its contents. Husserl emphasizes that it is "only through" the act that its contents are united.[15] Their wholeness is therefore peculiar to that of a "psychical whole." As parts of this whole the contents are described as "embraced," "surrounded," and even "captured" by the act. These descriptions clarify the meaning of "unifying interest" and Husserl sometimes conjoins them with the latter characterization.[16] He also describes this embrace of the contents or "foundations" (*Fundamente*) as "intentional" and as a case of "intentional inexistence."[17] The former term is essentially synonymous with those describing the act as being "trained" or "directed" upon its contents. His characterization of the act as "inexistent" reiterates and emphasizes its externality to its contents.[18]

Sixth, the totalizing act is able to unite its contents only insofar as it is an "act of noticing" (*bemerken*). In Husserl's initial descriptions of it, this noticing function seems to compromise the unity of the act, and its precise relation to the contents remains unclear. He writes that the "act of noticing" is "simultaneously with and in" this concentrated unifying interest, in the manner of "mutual pervasion" (*Durchdringung*) "peculiar" to psychical acts.[19] This description in general, and that of "simultaneity" in particular, indicates that these two functions are intimately related. Yet Husserl does not clarify whether or not they are facets of one indivisible act.

This act of noticing "lifts the contents out," but are these the individual contents successively drawn into relief by the act of noticing which, simultaneously, allows their field to recede as unnoticed? Or are they contents *qua* completed "totality" (*Inbegriff*) which are lifted out as the object of attention? The latter interpretation seems correct, for Husserl writes that it is "the representation of the multiplicity (*Vielheit*) or the totality" which is the object of this act.

He uses very similar words in a later passage when he writes that a "totality comes into being" (*entstehen*) in virtue of an "undivided interest" and, simultaneously, "in and with it" an "undivided noticing" which "lifts out" and "embraces different contents for themselves."[20] What are "lifted out" here are clearly the individual contents. While they are simultaneously "embraced" for themselves, it remains unclear whether this embrace is simply of an individual entity lifted out, or whether the latter is embraced as a part of an existing totality of contents.

In the last paragraph of *BZ* Husserl writes that "in all cases where discrete contents are thought together, i.e., in a totality, there is present (*vorhanden*) one and the same, constantly uniform act of collecting interest and noticing." The totality is further clarified here as the being "thought-together" of discrete entities. This act of "collecting interest and noticing" thus "separates each of the individual contents 'for itself' (i.e., as noticed 'for itself'), and simultaneously holds it together, unified, with the others."[21]

The unity of the act is beyond question in this passage. Its functions of noticing and lifting out each content, and of simultaneously holding it together with those previously noticed and lifted out, are two facets of the same act. They are not two acts.

The simultaneity of these two distinct moments of the act conveys the temporality of totalizing. That "unifying interest" which is the totalizing act must precede the noticing and lifting out of particular entities, logically as well as temporally. If it did not, the first such entity would never be lifted out. While this interest motivates the initial lifting out, the latter enacts this interest. Without this governing interest, these acts cease, but without these acts this interest remains only an intention. As the totalizing process unfolds, each act of adding an entity involves an awareness that this entity is now being united with a "totality" of entities which have already been successively united with one another. Conversely, there is an awareness that this existing totality is being expanded by another member. This may also be an awareness that "this is not the last member to be added," and thus an anticipation of further additions to the totality.[22]

Husserl acknowledges that this act is not anonymous but has a subject. It follows that this subject is aware that the totalizing process was initiated and sustained for a given duration simply because it had an interest in doing so and so willed it. The encompassing self-awareness of the totalizing subject is therefore teleological. The totalizing process is spontaneously set in motion, motivated only by the interest of the subject in gathering entities together. This overarching interest, present in noticing each content, provides the *telos* of each particular act. The latter is the means by which that end is achieved. Insofar as these are moments of the act of a subject, the latter determines both its end and the means by which it will be realized. This end and means are then also intimately related in their "mutual pervasion" of one another.

In sum, Husserl characterizes the totalizing act as that of an agent who is spontaneous and free. It is motivated by its own interest and therefore sets its *telos* (unifying contents) and the means to reach it (noticing each content separately while adding it to the existing collection).[23] It is self-aware over time, it is essentially external to the contents upon which its interest is concentrated. It is a fundamental unity.

The totalizing act *is* the psychical whole. The latter is convertible with this act and is not, therefore, a content in any sense of the word. The only contents of a totality are the *relata*. If, for instance, there is a totality of a *relatum* and another *relatum*, these are conjoined only in virtue of being thought together. Apart from them, there is only the totalizing act, and apart from it they are entirely unrelated. The "totality" *is*, therefore, the totalizing act.

The Totalizing Act as Totality

Husserl's characterization of the totalizing act as connected or "attached" to the contents suggests a certain externality of the former to the latter. The act connects the individual contents to one another, yet remains essentially non-contentual itself. This essential hiatus between act and contents is emphasized by Husserl's descriptions of the totality as a "psychical" whole, and the totalizing act as "intentionally inexistent."

Husserl's metaphors, describing in most "intimate" terms the relation of the totalizing act to its contents, may obscure this fundamental distinction. The act, he writes, "seizes" or "captures" its contents. It "surrounds" and even "embraces" them. These images may suggest that the act in some sense becomes captive to the contents in turn. It may seem that its spontaneous and arbitrary freedom is circumscribed and choked by the reciprocal "embrace" of its contents. In fact, however, the contents neither ensnare the act, thereby closing the "distance" between them, nor do they requite its embrace, thereby rendering it indistinguishable from them.

Husserl's rigorous distinction of the totalizing act and its contents by no means entails the conclusion that the act has no affect upon its contents. On the contrary, what were absolutely disparate entities become members of a "totality" in its psychical embrace. And its motivating interest and unifying activity are expressions of the unity of the act itself. Other qualities of the totalizing act directly affect its contents as well.

One such quality, "arbitrariness," expresses itself in prescribing the unlimited range of entities that may be collected. The fact that the act may unite any content with any other is not preclusive of the possibility of pattern in the contents. A thought may well unite "several specific trees," as Husserl suggests, or, "the sun, the moon, the earth, and Mars." But it may just as easily embrace "an angel, the moon, and Italy." Husserl's point is that such groupings are entirely at the discretion of the totalizing subject regardless of what they may consist. They "depend in no way on the nature of the individual contents."[24] What the latter are in particular explains not one whit why these rather than those were singled out by the totalizing subject. The constituency of the given totalities depends solely on the arbitrary fiat of this agent.

The arbitrary capacity to think anything together with anything else, is one of which the subject must be aware virtually from the outset. While it need not be a thematic awareness, it conditions and informs the collection of specific entities at the most concrete level. The child who collects seashells is aware that it could collect insects, or anything else it liked, if it were "interested" in doing so and had the capacity. Normally, contents thought together are united because of a special interest in bringing them in particular together. This does not nullify the penumbral awareness that "anything else" could have been thought together with "anything else." In this respect, thoughts are much less limited than capacities for collecting material objects.

The totalizing act is not merely the absolute arbiter of which entities become members of its exclusive totalities. It is the totalizing element in the absence of which there remain only solitary and wholly unrelated individuals. The "totality" of contents *is* their being "thought-together." Of this Husserl provides further, and unmistakable, confirmation when he writes that "if we execute this act, then we would naturally seek in vain for a relation or connection (*Beziehung oder Verbindung*) in the content of the representation which it embraces." The act which connects contents is not itself a content. This, Husserl implies, is obvious, and he emphasizes it in the following sentence which begins, "The contents are here united 'only just' by the act. . . ."[25] The insistence on the totalizing act as the sole unifying source is the simultaneous acknowledgement of its absence and that of any other connection among the contents. Husserl reiterates and clarifies this point in an approving citation of John Stuart Mill's definition of "relation." Mill maintained that the contents are unified in virtue of the fact that they are "just merely thought'-together', i.e., thought as a totality."[26] The absence of the "thinking-together" act among the contents does not imply that it is evanescent. It is *also* a "holding-together" act, Husserl continues, and is therefore eminently extant. It is absent only, he adds, "in the contents."[27] Husserl emphasizes the unifying power of this act when he writes in another passage that to "think-together" the contents is to think them "in *one* act."[28] The unity of the contents is tantamount to the unity and singularity of the act. Husserl insists once more that it is only this act, embracing contents in "a *psychical* whole," which "amounts to their relation."[29] The totalizing act is in no manner "objectivated," then, other than in reflection. On the contrary, its contents are "act-ivated" as much as is possible at this point.

The totalizing act provides the basis Husserl needs to speak of the collective connection.[30] These are equivalent but not identical. Husserl repeats the reason for this at a number of points. The collective connection, he stipulates, can "be noticed . . . only by means of reflection on the psychical act, in virtue of which the totality comes about."[31] The collective connection is simply the totalizing act "objectified" by a reflective act directed upon it.

The transition from totalizing to reflecting upon this totalizing is not mysterious. The agent is perfectly aware of the activity underway and of itself as the thinker whose thought embraces the contents. The condition, then, for this transition to reflection is simply that of self-awareness. The motivation to reflect comes not from a perception of contents as "together," and the subsequent query of how they came to be (a query possible only in the context of some sort of ignorance of oneself as the totalizing subject). The objectification of this act in reflection is entirely other than its "contentualization" which construes it as being a quasi-content among contents.

While Husserl does not characterize the collective connection differently than the totalizing act, he treats it within the broader context of relations which unite various wholes. He is firmly persuaded that the former explain the latter in every case. The classes of wholes are circumscribed by reflection on the "well-characterized manner of connection of the parts."[32] The salient feature of the "psychical whole" is that its "manner of connection" is missing from among its parts.

Husserl employs the well-honed distinction between acts and contents to classify this particular relation and distinguish it from its opposite. His categories of "physical" (or "content") and "psychical" (or "act") relations essentially recapitulate insights already gained (i.e., that psychical acts are in principal external to their contents). Physical relations belong to the contents of the representation in the same sense as do the *relata* on which they "rest" (though it is still by means of these relations that the whole is characterized and classified),[33] and they might be appropriately termed "content relations." They are "intuitably contained" among the contents of totalities as "part-phenomena" and may be "separately noticed" (*bemerken*). "Psychical" relations are noticed, on the contrary, "only in the psychical act."[34]

The intuitability of content relations does not necessarily entail their being immediately obvious. Analysis of their wholes may well be necessary, Husserl argues, in order to notice these "binding" relations. The physical parts of a rose, in his example, are analytically distinguished, then the properties of the parts. The procedure Husserl describes is nothing but the totalizing act in reverse. Each part is "lifted out" by a particular noticing and held together with all the other parts already noticed. Analysis completed, what remains is a "*totality*" of parts formerly comprising the whole. A simultaneous reflection on this "whole" reveals its "binding relations" as separately and specifically determined phenomena of the representation. These relations, he writes, give themselves as "the more" *in contrast to* the "simple totality" which seems to "capture and not connect the parts."[35]

Husserl's analysis of the rose is important not merely because it clarifies the essential difference between the psychical and physical relations. While such clarification was doubtless his intent, it is effected by employing the totalizing act beyond its natural domain. The content relation of the pre-given whole called "rose" is articulated only through a sort of totalizing in reverse. This is the first of many instances in which Husserl employs the totalizing act as the means by which "non-totalized" sensible wholes are understood or appropriated by the psyche.

Husserl clearly contrasts "pure" or "simple" totalities and what might be termed "adulterated" totalities, those in which the connections are intuitably "something more" than the parts. More significantly, he contrasts these "binding relations" with the "totality" or "psychical whole." The latter is clearly tantamount, in his view, to the collective connection, for it functions (i.e., "captures") as a *relation*. The metaphor of capture is an apt characterization of the psychical act's relation to its contents. The group of arbitrarily apprehended or captured persons has nothing in common than the fact of their capture and common captor. Since there is nothing "more" in the simple totality than the parts themselves, the collective connection appears "as the case of *relationlessness*" when compared to those of content relations. In the absence of physical relations, contents can only appear to be "'disconnected' or 'relationless'." And these contents are none other than those "just simply thought-'together,' i.e., thought as totality." Husserl hastens to add that these contents are not "actually" bereft of connection. His hyperbole only articulates the absence of "noticeable unification" in the case of the "holding-together psychical act."[36]

Husserl also describes the psychical act or relation as a "psychical phenomenon" which embraces its contents "intentionally." He does so in the context of describing "relations" (which would seem to refer to the collective connection), but he speaks also of acts. In any case, these are members of a class of psychical phenomena which embrace their contents intentionally (e.g., acts of noticing, willing, etc., embracing the noticed, the willed, etc.) Husserl gives a more intimate characterization of intentionality here than the famous one of simple "directedness." He compares the manner in which the physical relation of similarity "embraces" its contents with that of any "intentionally inexistent" relation. They are, he writes, entirely different manners of "inclusions" (*Einschlüsse*). "Intentional" relations are virtually "inexistent" only in that they are not "contentual."

The collective connection belongs, Husserl writes, "to a new class of relations well distinguished from all others," one that is "peculiar." Its totality is a "whole of a particular kind" united by relations "exclusively characteristic" of it.[37] Indeed, Husserl regards the collective connection as the most fundamental of relations. It is "an indispensable psychological precondition for every relation and connection whatsoever."[38] Since Husserl maintains that all wholes are explicable only with reference to their connections, it follows that the totalizing act qua collective connection is the "psychological pre-condition" for every whole. It becomes such in Husserl's epistemology because it is not merely the "precondition" for the psychical whole. It *is* this whole of otherwise disparate and unrelated entities. In its absence their relatedness is wholly unintelligible.

This will become more evident in the following chapter. There it will be seen that the concept of the collective connection, and those of the number concepts deriving from it (and from that of something abstracted from the representing act), *share* the "externality" to contents of the acts from which they are abstracted and derive. Their experiencing in straightforward association with contents is entirely due to their "attachment" in a "non-literal" sense to the latter by the psyche. These concrete collections do not consequently serve as the primary abstractive bases for these concepts. The necessity of the latter being "attached" witnesses to their exclusive origin in totalizing acts and to these as the undisputed origins of totalities themselves.

It will become clear as well that the concepts the psyche derives autoabstractively refer to it and to its totalizing capacities alone. The ubiquity of the totalizing act is felt at virtually every turn of Husserl's generation of the number concepts. It provides the abstractive base, it functions as the abstracter of its own self-referential concept, it applies the latter to concrete collections and abstracts them from these as well. And this mode of the psyche is also the central mechanism by which concepts of other entities are abstracted. In gathering characteristics of objects exhibiting the content relation of similarity, it generates the abstractive fundament for the given concept.

This act is then the precondition for the articulation of this *content* relation as well, and of the concept which unites the particular characteristics perceived as similar. This case is not an anomaly, but simply a demonstration of Husserl's contention that the totalizing psyche is the "indispensable psychological precondition for every relation and connection whatsoever." The analysis of the rose is but one demonstration of this. This pre-given, physical whole was virtually dismantled and reconstructed by and through the totalizing act. Again, its content relation was articulated only after this had been accomplished.

Virtually from its inception as the pivotal player in Husserl's conceptual mathematical project, the totalizing act or mode of the psyche ranges far beyond the parameters of this endeavor. It is the source of currents which would ultimately bear Husserl along on an intellectual adventure he could not have imagined in 1887 at Halle.

NOTES

1. Edmund Husserl, *Über den Begriff der Zahl: Psychologische Analysen* (Halle a.S.: Heynemannische Buchdruckerei/F. Beyer, 1887). Republished in *Philosophie der Arithmetik* (1890—1901) edited by Lothar Eley, *Husserliana*, Band XII (The Hague: Martinus Nijhoff, 1970). All references to this article (hereafter *BZ*) are to the Nijhoff edition See *BZ* 295, 2—7.
2. *Ibid.*, 291, 13—17. Morris Kline [*Mathematics: The Loss of Certainty* (New York: Oxford University Press, 1980)] asserts that irrationals were not so problematic by this time because they could be represented as points on a line. It was probably this need for geometrical props to which Husserl objected. See p. 153f.
3. *Ibid.*, 291; 294, 20—37.
4. *Ibid.*, 294, 25—33.
5. *Ibid.*, 292, 32—293, 4.
6. *Ibid.*, 295, 13—30.
7. Dallas Willard discusses this hierarchy and succinctly describes the "more important . . . 'contents of consciousenss,' according to Husserl's 'initial view.' " [*Logic and the Objectivity of Knowledge* (Athens: Ohio University Press, 1984), pp. 36—38 (hereafter *LOK*)]. See also pp. 29—34, where he presents Husserl's conceptual analysis against its historical backdrop from Hume onward. Theodore de Boer also discusses this topic in some detail. See Chapter One of his *The Development of Husserl's Thought*, translated by Theodore Plantinga (The Hague: Martinus Nijhoff, 1978), especially pp. 4—9 and 40—50 (hereafter *DHT*).
8. Husserl emphasizes that every "collection" presumes a "colligating act" (*kolligieren*). *BZ* 305, 6—10. See also Edmund Husserl, *Philosophie der Arithmetik. Psychologische und logische Untersuchungen, Erster Band* (Halle-Saale, 1891). Republished in *Philosophie der Arithmetik* (1890—1901) edited by Lothar Eley, *Husserliana*, Band XII (The Hague: Martinus Nijhoff, 1970). All references to this work (hereafter *PA*) are to the Nijhoff edition. See *PA* 25, 16—19.
9. *BZ* 316, 30/*PA* 45, 13. In the context of this reference, Husserl substitutes, at *PA* 45, 9, "It is indubitable" (*unzweifelhaft*) for, "it is certainly correct" (*richtig*), at *BZ* 316, 27, and "separated" at *PA* 45, 14, for "discrete" at *BZ* 316, 32. Since "separated" is used in *BZ* (at 317, 4, and reproduced at *PA* 45, 24), these alterations perhaps indicate a harmonizing of *BZ* with itself more than anything else. However this may be, *PA* has no less of an interest in indicating the psychic fundament of such notions. See 330, 31—37/69, 15—21; 316, 31/45, 14.
10. *Ibid.*, 316, 30/45, 13.
11. *Ibid.*, 330, 28/69, 11. cf., *BZ* 316, 34/*PA* 45, 17—18. While the point is essentially the same in each of the latter parallel passages, that in *PA* might be interpreted as placing more emphasis on the act. In the *PA* passage it not only unifies the contents but is itself "unified." It not only "connects" but "encompasses" them. More than simply being directed upon them, it "surrounds" them. However this "directedness" is certainly not dropped from *PA* as a characterization of the act (see *PA* 30, 21/*BZ* 308, 36). It should not be inferred, either, that the characterizations of it in this passage of *PA* are

missing at other points from *BZ* (see *BZ* 329, 30/*PA* 68, 15; *BZ* 330, 32/*PA* 69, 16; *BZ* 332, 8/*PA* 72, 25). Franz Brentano uses the verb "*richten*" to describe this psychical activity as well (see his *Psychologie vom empirischen Standpunkt, Erster Band*, ed., Kraus, Leipzig, 1929, p. 124f.).

12. *Ibid.*, 330, 29/69, 12.
13. *Ibid.*, 316, 34/45, 17—18. In *PA*, Husserl substitutes "connecting" (*verknüpfen*) for "unifying" (*einigen*) in *BZ* as a modifier of "interest." Husserl uses the word "connect" also in *BZ* to characterize the nature of the relation of the act to the contents. This additional use of the term may imply that the relation of the act to the contents, and the consequent relation of the contents to one another, is of a piece. There are other instances where a characteristic of the act is expressed in its contents, e.g., Husserl implies that the unity of the contents is a function of the unity of the act (cf., *BZ* 334, 3/*PA* 74, 19).
14. *Ibid.*, 317, 4/45, 23. The sentences are identical with the exception that, in *PA*, the "complex" is inserted before "psychical act" (*BZ* 317, 5/*PA* 45, 25).
15. *BZ* 330, 35/*PA* 69, 19.
16. *Ibid.*, 332, n. 1, 4/73, n. 1, 4. This whole is "psychical" because (Husserl writes, endorsing the view of J. S. Mill) its "parts" are united by a "psychical act thinking (them) together" (*zusammendenken*). Compare this passage to that at *BZ* 330, 29/*PA* 69, 12. Husserl does explicitly warn against the use of the metaphor of "whole/parts" because it implies more of a unity among the contents of this act than he wishes to convey that there is (see *BZ* 335, 7—10/*PA* 77, 16—21; the latter passage makes essentially the same point, albeit paraphrased). 329, 30/68, 15; 330, 32/69, 6; 332, 8/72, 25; 331, 25—6/72, 3—4.
17. *Ibid.*, 329, 323/68, 17—18. Husserl's use of the term, "foundation" (*Fundament*) as an apparent synonym for the contents collected, less likely indicates that the totalizing process is radically dependent on these (it certainly requires contents in principle, but is by no means determined by the particular natures of such) than it does that this process presumes individual representing acts of such entities, prior to any thought of their being totalized (at any rate, this is his view of the matter by the time of *PA*).
18. With regard to the use of the term, "intentional" as indicating relatedness or directedness to a content, de Boer notes (*DHT*, p. 9) that, in the corpus of Thomas Aquinas, "*intentio*," in the cognitive order, refers to the specific "objective" (as opposed to "real") mode of being of the object in the mind of the knower. Only in the connative order does it refer to a "striving after" or an "impulse" for Aquinas ("*Intentio est proprie actus Voluntatis*."). According to de Boer, Brentano adapted this meaning with reference to cognitive acts, as Husserl does as well.
19. *BZ* 300, 29/*PA* 19, 34. It is not entirely clear what Husserl means to convey by the use of "peculiar" (*eigen*) here, since he also speaks of the spatial extension and color, and the latter and intensity, of the object as "connected" (*verknüpfen*) in "mutual pervasion" (*gegenseitiger Durchdringung*). cf., *BZ* 300, 26—9/*PA* 19, 31—4.
20. *Ibid.*, 333, 28—334, 3/74, 7—19. There are minor changes in the *PA* rendition of this passage. Perhaps the most significant is that lines 10—13 are not emphasized by Husserl in *PA* as they are in *BZ*. *PA* also substitutes "grasp" in this sentence for "notice" in *BZ*. Perhaps this indicates an "activation" of the act, at least in this passage.
21. *Ibid.*, 337, 25—32. This text is absent from *PA* because the parallel texts diverge after *BZ* 337, 9, and *PA* 82, 8. The remaining few pages of *PA* constitute an expansion of *BZ*.
22. The outlines of Husserl's detailed investigations of inner temporality are evident here (and are developed much further in his article of 1894, "Psychological Studies for an Elementary Logic.")

The Totalizing Act: Key to Husserl's Early Philosophy 29

23. cf., *BZ* 308, 30—6/*PA* 30, 15—21, where Husserl also emphasizes the noticing facet of the act.
24. *BZ* 298, 26—299, 2/*PA* 16, 16—25. The passages are essentially the same, however *PA* includes the phrase, "And accordingly also enumerated" (*PA* 16, 20/*BZ* 298, 29), and substitutes "multiplicity" (*Vielheit*) at line 23 for "plurality" (*Mehrheit*) at line 32 of *BZ*.
25. *BZ* 330, 34—6/69, 18—20. The phrase, "*eben nur*," can be rendered other than "only just," but regardless of translation, the meaning seems to be that the act is the sole unifying source and force. This locution then emphasizes the absence of the totalizing act among its contents.
26. *Ibid.*, 332, 29—30/73, 3—4. This phrase, "*eben bloss*," functions similarly to "*eben nur*" in the preceding note. cf., *BZ* 334, 3, and *PA* 74, 18. In the latter *PA* passage, Husserl drops the quotation marks surrounding "together," and "one" is not emphasized as it is in its antecedent *BZ* version. This might be taken to suggest greater cohesion among the contents and, by implication, a more powerful act. Such a situation does not then necessitate an emphasis on the unity or singularity of the latter. However the fact that the quotation marks are extant in other places in *PA* (cf., *PA* 73, 3—4) prevents any sustained interpretation of this sort.
27. *Ibid.*, 332, 33/73, 7. Husserl actually speaks here of the absence of the "unification" (*Einigung*). Within the context, he might as well have referred to the "holding-together psychical *act*" as he does in the preceding sentence.
28. *Ibid.*, 334, 3/74, 19.
29. *Ibid.*, 332, n. 1, 3—5/73, n. 1, 3—5. These two passages are virtually identical, with the exception of the substitution in *PA* of "primary relations" for "content relations" in *BZ*. At *PA* 70, n. 1, Hussel comments on his avoidance of the term, "physical phenomenon," and emphasizes that "intentional inexistence" is the distinguishing mark of "psychical" relations when compared to "primary relations." This was also the main criterion for Brentano in distinguishing "physical" and "psychical" phenomena. The term "physical phenomenon" refers then to "primary" absolute contents, not to their "abstract moments" (which [even "physical"] relations are).
30. *BZ* 301, 19—21/*PA* 20, 26—8.
31. *Ibid.*, 333, 32—5/74, 10—13. cf., *BZ* 330, 34—6/*PA* 69, 18—20 for the characterization of this reflection as "peculiar" (*besonder*). At line 20 of *BZ* 333/*PA* 73, 7, the connection is also called the "collective unification" (*Einigung*).
32. *Ibid.*, 300, 34—8/20, 3—7.
33. *Ibid.*, 329, 24—30/68, 9—15.
34. *Ibid.*, 331, 6—26/69, 29—33; 71, 19—72, 4. Husserl interpolates the *BZ* text in *PA* at *BZ* 331, 10, and between *PA* 69, 33, and *PA* 71, 22. *PA* 69, 29—33, paralleling *BZ* 331, 6—10, makes essentially the same point with somewhat different language, including the usual substitution of "primary" for "physical" or "content" with regard to relations. For this reason, *BZ* 331, 10—11 is omitted by *PA*. In this sentence Husserl cites "content relations" as a viable synonym for "physical relations." Similarly, with the exception of a couple of minor alterations in *PA*, *BZ* 331, 12—26 is essentially reproduced at *PA* 71, 24—72, 4.
35. *Ibid.*, 331, 29—332, 9/72, 5—25. Even concepts of "physical" relations are gained via "reflection" (cf., *BZ* 300, 22—4/*PA* 19, 27—30). This reflection is, however, not reflexive, since it is directed upon contents and not acts.
36. *Ibid.*, 332, 17—33/72, 30—73, 7. It is striking that *PA* omits *BZ* 332, 12b—16. Since *BZ* 332, 25-33/*PA* 72, 38—73, 7, more or less amplifies it, Husserl may have regarded it as redundant for *PA*. The passage has been used as one of the foundations for the claim that some sort of contentual status is suggested here for this "unity"

(*Einheit*). Perhaps Husserl himself regarded it, four years later, as sufficiently ambiguous and suggestive of a "physical" unity among the contents (which he propounded neither in *BZ* nor in *PA*) as to warrant its deletion from *PA*.

37. *BZ* 328, 13—19/*PA* 66, 5—11. Husserl amplifies the point in *PA* by appealing to "inner experience" (*Erfahrung*), and by insisting that the collective connection is neither to be "dissolved into" nor defined in terms of any other relations.

38. *Ibid.*, 334, 13—14/75, 26—28.

CHAPTER II

The Concept of the Totalizing Act as Collective Connection: Progenitor of Number

THE AUTO-ABSTRACTION OF THE CONCEPT OF COLLECTIVE CONNECTION

The collective connection is discerned in the "thinking-together act." It *is* the latter objectified in reflection. This reflective act instigates the debut of the collective connection as a relation among relations, albeit a "peculiar" relation binding its parts into a peculiar whole.

Husserl also describes a reflective act which is an *abstracting* act directed upon the totalizing act. "We gain the abstract representation of the collective connection in reflection on that elementary act of 'lifting-out' interest and noticing. . . ."[1]

The ambiguity in Husserl's usage is evident in this passage. While he often states that the collective connection is seen only in the psychical act, he sometimes conflates the two as he apparently does in this passage. The "elementary act" of which he speaks here *is*, qua object of reflection, the collective connection.

Presumably, the "abstract representation" of the collective connection would be gained by *another* reflective act directed upon the collective connection. Or, since any reflective act directed on the "elementary act" transforms it into the collective connection, perhaps the abstractive act requires no prior reflective act mediating the totalizing act to it. Husserl neither spells this out nor makes these distinctions. While the abstracting act perhaps need not be predicated on the reflective collective connection-discerning act, the two may be distinguished with regard to their respective intents. The intent to reflect on the "thinking-together act" need not be more than that. Conversely, the abstractive act need not be a reflection *per sé* on the collective connection. It *is* an abstractive reflection motivated by the intent to gain the abstract representation of this connection.

Husserl speaks of the abstractive act as directed upon the elementary *act*. While there is precedent in *BZ* for conceptual abstraction on the basis of only one instance (e.g., the concept of something), Husserl's general doctrine of abstraction stipulates that more than one instance is required.

According to this doctrine, "concrete phenomena" form the "basis (*Grundlage*) for abstractive acts which "rest" (*beruhen*) on these "irreducible facts." Their characterization is the task of "psychological description."[2] The process of abstraction and that which is abstracted is clearly derived from, and predicated upon, particulars. The concrete is the contrapuntal correlate of the abstract, but the concrete need not be sensuous, only "factual." The "abstract representation" and the "universal representation" are abstracted from the concrete.[3]

Husserl describes the process of abstraction as "moving from" (*ausgehen*) concrete phenomena to general concepts,[4] and it is this movement that must be clarified. Similarity (*Ähnlichkeit*) is a content relation that is "intuitable." Identity (*Gleichheit*) is another.[5]

Concrete phenomena exhibit similarities, or even virtual identities, and it is these to which attention is directed as the relational bases for abstraction. Since Husserl regards the concrete phenomena exhibiting these content relations as the bases for abstraction, there are then really two foundations. The basis of the abstract representation is the "constant constituent attended-to at all times,"[6] that which is always "similarly present."[7]

Husserl's doctrine of abstraction is essentially reliant on acts of attention (and inattention). To abstract is simply to "disregard" the "characteristics (*Merkmale*) that are different" in the cases inspected and to "hold fast" to those "common" (*gemeinsam*) to all.[8] This "attending" (*achten*)[9] and holding fast is an act of "comparison" (*Vergleichung*).[10] Husserl characterizes this attention to constants in the same manner as he did the "lifting out" of individual entities to be added to the collection. The constant characteristics become "objects of an interest which 'singles them out'" and "notices them for themselves."[11] Since attention is directed toward what is common, no "special interest" is paid to the "peculiarities of the contents." Abstraction seems to be defined by this inattention, yet the latter is clearly a function of attention.[12]

It is by means of this attending to one characteristic exhibited by a number of cases (and disregarding its variations case to case) that the "concept" (*Begriff*) of the characteristic is "formed" (*bilden*) and "comes into being" (*entstehen*). It is those characteristics common to all cases which "constitute" the general concept.[13]

Husserl's abstractive doctrine is a further instance in which he employs the totalizing mode beyond its original territory. It is not that it collects particular characteristics which are similar, but that it attends to a collection of these and abstracts that which "unites" them (the "universal representation"). The totalizing act *articulates* the essential meaning of the *content* relation of similarity as it did that binding the rose.

It must be emphasized that Husserl's concept is not a particular representing a collection of particulars observed in a number of cases. In this sense, his doctrine is not that of classical empiricism or nominalism. This can be seen clearly in the application of his formula for abstraction. Particular colors, for instance, are seen to be similar in a number of instances. Slight variations in shade may be detected but are disregarded by attention to "the color in all cases." This attention is to "*nothing* 'in particular.'" To put it in this way is not to pronounce on the ontic status of the "non-particular" or "general." The concept is simply an "abstraction," the result of an abstracting process.

Is, then, the abstract representation of the collective connection predicated on a comparison of a number of cases or on just one? In instances of other wholes and their particular connections, Husserl does assay a number of cases.[14] This seems to be the practice indicated as well when he writes that the abstraction of the concept of the collective connection "is possible because . . . in all cases . . . there is present (*vorhanden*) one and the same, constantly uniform (*gleichartig*), act of collecting interest and noticing. . . ."[15]

The constant in each psychical whole is the particular totalizing act. But what are the differences which distinguish one totalizing act from another? It is apparent that the totalizing act objectified as the connection is essentially different from other abstractive bases.

The collective connection derives from a function of its subject directed upon an act of the same subject. This "foundation" is one to which the subject is then intimately and uniquely privy. Based on the content of this self-awareness, insight may be gained as to how the abstraction of the concept of the collective connection might be unique. What then is "peculiar" to individual collective connections which must be disregarded and what remains constant and common to all?

It cannot be the particular contents which distinguish one collective connection from another, since what is collected is simply a function of the interest of the given act. Might, then, given connections be distinguished by their particular interests?

The fact that *each* act *could* have collected altogether different things, and had a correlative interest in doing so, seems to imply a negative answer. Each particular totalizing agent may well be aware that it can randomly "think together" a succession of "whatever comes to mind." If this is the defining awareness *in each particular case*, then it would seem that it is not the incidental particular "interests" which distinguish collective connections. There may be an interest in uniting even abstract "whatevers" with one another. However, this is most fundamentally not an interest but a capacity of the agent.

If each totalizing act involves an awareness of its capacity to unite anything with anything else — i.e., *it* is an act which unites anything with anything else — what distinguishes one from another? Since each act proceeds for a given duration, it seems that respective temporal coordinates are the only distinguishing features. These alone define the "particularity" of each act. Due to the unique character of this abstractive basis, the abstraction of its concept hardly seems to require reflection on a number of cases. If reflection is involved, it but articulates an awareness of a capacity which is essentially formalizing and self-formalizing.

The concept of the totalizing act *qua* collective connection is the product of an auto-abstractive act of the psyche. Arguably, this totalizing act *is* virtually formal. It therefore renders an assay of a number of its instances, in order to abstract its concept from them, superfluous. Whether or not this was Husserl's view is unclear.

It is clear that the concept of collective connection is that of the totalizing act. This concept *is* the *conceptual* totality or multiplicity (*Vielheit*) in much the same way as the totalizing act *is* the totality on the plane of particulars. It is on the basis of this concept that Husserl generates the determinate number concepts. The lineage of these concepts can therefore be traced to the totalizing act. It was Husserl's great hope that the highest reaches of pure mathematics might be demonstrated to be the progeny of this primordial parent as well.

NUMBER CONCEPTS: PROGENY OF THE TOTALIZING ACT

Having gained the abstract representation of the collective connection, it is "by means of it," Husserl writes, that "we form (*bilden*) the universal concept of multiplicity (*Vielheit*) as a whole, which connects parts in a simple connective manner." The concept of the collective connection is explanatory of the abstract whole in the same way as the totalizing act is explanatory of the concrete psychical whole.[16]

While the concept of collective connection is necessary for the formation of this concept, it is not sufficient. Even on this abstract or conceptual plane, the connection must connect something. Precisely, it must connect the concepts of "something."

Husserl asserts that it is obvious that the "concept of something" owes its "origin" (*Entstehung*) to reflection on the "psychical act of representing." The content of this act is "just any determinate object." He emphasizes that this concept is not won through comparison of contents of either physical or psychical representations. It does not originate by comparing a variety of cases exhibiting a similarity called "something."[17] The analogy of this analysis to that suggested for the concept of the collective connection is obvious. Both the totalizing and representing acts may direct themselves to anything. While Husserl asserts that the concept of something originates in reflection on the representing act, it is not clear that this is necessary any more than in the case of the totalizing act. It is a common experience to think of "everything" but what one is supposed to. This awareness, though not identical to the recognition that one can represent anything one chooses, is not far from it. This act seems to be a necessary condition or even foundation of the totalizing act. The latter may add "anything" and "anything" only because individual representing acts may represent "anything" (and, at a given time, "something").

The concept of multiplicity is then comprised of those of collective connection and something conjoined in the manner of " 'something,' 'and' 'something,' 'and' 'something,' etc." Essential to this concept, Husserl writes, is "an indeterminacy" (*Unbestimmtheit*), and it is "etc." which signifies this characteristic of the series.

The concept of multiplicity, like its concrete counterparts, involves a group of indeterminately many entities. However, this does not entail an infinite progress in thinking the concept. "De facto" in thinking a "multiplicity," Husserl observes, a "limit" (*Begrenzung*) is reached. Insofar as thinkers of the concept realize it to be but an "arbitrary" (*willkürlich*) limit, it signifies very little conceptually. Husserl seems to imply that this limit is indicative of the finite incapacity to grasp more than a small multitude of entities individually and as members of a determinate group. This limit, then, appears to be set by the arbitrariness of psychic capacity rather than by that of will.

Husserl is more interested, however, in the "improvement" (*Hebung*) or determination of this indeterminate series that *is* subject to psychic intention. The psyche is responsible for the "determined concepts of multiplicity" (*Vielheitsbegriffe*) or "numbers" (*Zahlen*). Husserl's procedure recapitulates the classical conquest of "*to āpeiron*" by "*pēras*." These determinate concepts are "specializations" of the concept of indeterminate multiplicity, distinguished from one another only by their respective limits imposed by psychical fiat. A number name is then given to newly-determined concepts.[18]

The *conceptual* nature of numbers *qua* deteminate conceptual multiplicities must be emphasized here as it is by Husserl himself. Their nature as such may be obscured by the incorrect appraisal of his statement that "the designation of numbers as pure mental creations (*Schöpfungen*) of an inner intuition involves an exaggèration and distortion of the true state of affairs." It is clear from the context of this assertion that Husserl is less concerned to dissociate numbers from "mind," than he is from the notion of "creations" *in the sense of* "new absolute contents" which could be found again in "space" or the "external world." Numbers, Husserl insists, are in no sense spatio-temporal deposits of psychical acts. They *are* psychical creations insofar as they are the "results" of activities "exercised" upon concrete or particular contents.[19]

Number concepts are generated by psychical acts from concepts of psychical acts. They trace their descent to psychical acts, and primarily to totalizing acts, via their constituent concepts of something and collective connection. They also stand in the relation to the act of product to producer. Psychical activities "produce" (*schaffen*) these "peculiar relational concepts," Husserl writes, which can be "produced (*erzeugen*) again and again." Yet so dependent on the act are these concepts that they "in no way can be found completed (*fertig*) anywhere."[20]

As a consequence of these intimate relations to psychical acts, number concepts are associated with whatever they number only in virtue of having been "attached" (*knüpfen*) to the latter. As "attachments" they indicate the psychical acts which attach them. Their status as attachments also reiterates that they are abstracted from, and products of, psychical acts which are themselves only "attached" to their contents.

The "Attachment" of Number Concepts: Index of the Totalizing Act

Husserl insists throughout *BZ* that concepts are not "part-phenomena" or "partial contents" of the particulars from which they are abstracted. A characteristic of a content, even the content relation of similarity uniting a number of recurrent cases of the characteristic, is incorrigibly particular. The concept of either the trait or the relation is emphatically not particular. It cannot be found intuitively as some aspect in or of concrete phenomena themselves. *What* then is the relation between it and such phenomena?

The answer to this question is motivated largely by Husserl's derivation of numerical concepts from those of reflectively objectified psychical acts. Psychical acts are, in principle, external to their contents, and the totalizing act is, therefore, external to its contents. Since all wholes are explained by Husserl in terms of their modes of connection, the totalizing act is explanatory of the "psychical whole" or totality. Thus, the concept of the latter must in some sense derive from that of the act. The collective connection is the totalizing act objectified by reflection, and its concept is that most fundamental to and explanatory of the concept of the totality.

The concept of the collective connection is gained through an abstracting act directed upon it. While its concept is fundamental to that of multiplicity, the latter also requires conceptual relata. These are produced by abstracting the concept of the representing act, since it may arbitrarily represent "anything" and therefore must represent "something." The latter concept is not gained through abstracting from objects but from the act which immediately formalizes an entire range of targets of thought. The concepts of collective connection and something are then conjoined to form an indeterminate conceptual series or multiplicity. Determining acts impose "limits" at each "something" of the series, thereby creating determinate conceptual multiplicities or totalities, and these are the number concepts. Husserl gives no further indication that abstraction of concepts is necessary. These determinate groupings are given the number names.

It would seem to follow that numbering or counting concrete entities would be the applying of these concepts to determinate groups. It is precisely the metaphor of "attaching" (*knüpfen*) which Husserl invokes to characterize the relationship between these "determinate concepts of multiplicity" or "numbers" and the concrete collections which exhibit them. This is the same term with which Husserl characterizes the relation of the totalizing act to its contents. Since number concepts are both abstracted from, and produced by, psychical acts, it is not a mere coincidence that this term is used of them as well. Husserl employs virtually the same wording in their case as in that of acts themselves. "We attach determinate numerical statements" to "sets of determinately given things." These in turn "fall under" the universal concept.[21]

This thesis of the externality of numerical concepts to concrete groups finds corroboration, albeit somewhat circumspect, at other points in *BZ*. Husserl observes, for instance, that the concept of something is related to concrete contents in "precisely the same manner" as the concept of number is to a "totality of concretely given contents."[22] Since the concept something is not derived from concrete contents but from the act in which they are represented, it could be inferred that this concept could also only be attached to concrete entities. Husserl asserts precisely that when he writes that the concept "something" belongs to concrete entities "only in the external (*äusserlich*) and non-literal (*uneigentlich*) manner" of "relative and negative attributes."[23] By implication, numerical concepts are also related to determinate concrete multiplicities falling under them in an "external" and "non-literal" manner. And this is so because they are essentially concepts of psychical acts in *BZ*.

This has not been evident to all commentators on *BZ* (or its sequel in the *Philosophy of Arithmetic*). It is often asserted that Husserl's account of the relation of number concepts to concrete collections is at best ambiguous, and at worst confused (and this either conceptually or semantically). Ironically, this view has perhaps been instigated more than anything else by the sentence continuing the passage cited concerning the "attachment" of "determinate numerical statements" to "concrete multiplicities." Husserl contends that since numerical concepts are attached to concrete phenomena falling under them, it is necessary "to proceed from" (*ausgehen*) these phenomena. The point of this procedure is to see how "with respect to" concrete collections the concept multiplicity and its determinate specifications "will be abstracted."[24]

It seems that Husserl is suddenly proferring the view that numerical concepts, and by implication those concepts constituting them, are not abstracted from the totalizing act *qua* collective connection but from its contents. What it means to proceed from concrete collections might be clarified by the assertion that the concepts are to be abstracted "with respect to them."[25] One might contend that the latter phrase does not state explicitly that concepts are abstracted *from* concrete collections. However, in another passage, Husserl speaks clearly of "abstraction from" these concrete phenomena.

Yet, the context of this locution throws this interpretation of it into serious question (i.e., that Husserl maintains that concrete collections are the abstractive bases of numerical concepts). Just after it Husserl writes that the first question we have to ask is that concerning the "*origin*" of the concepts in question. He seems to imply that this question has not been answered. But was it not answered in the preceding sentence which said that the concepts are abstracted "with respect to" concrete multiplicities? The next sentence states that these concrete phenomena "form the basis for the abstraction" of these concepts. The obvious question with reference to these two consecutive sentences is whether the origin of the concepts *is* their abstractive base in concrete phenomena.

A little later in this passage, there seems to be a question as to whether even the "proceeding from" concrete totalities *is* to be identified with the "abstracting from" them. Husserl himself asks how this is possible. "*Which* process of abstraction ought to yield it," he asks, and, most importantly, "what is that *from which* it (the concept) becomes abstracted?"[26] Given this last query, it is clear that Husserl is emphasizing that there is some question whether the concrete phenomena are that from which the abstraction is made, and whether "proceeding from" them is tantamount to "abstracting from" them.

Further, this passage begins with the description of the attachment of concepts to concrete phenomena. It seems to be the fact that the latter "fall under" these concepts which provides justification for "proceeding from" these phenomena. Unless this passage is a complete reversal of the position set forth rather clearly in *BZ*, and unless it is internally contradictory (reversing a position espoused in its opening lines) then it seems that its discussion of proceeding from the concrete phenomena must be interpreted as purely didactic. To employ the traditional distinction, Husserl is arguably detailing the *ordo cognoscendi*, not the *ordo essendi*. If these concepts are exhibited by concrete phenomena and abstracted from them, such is possible only in virtue of their having been "attached" to such phenomena. This is not to imply that the term "abstraction" here is entirely empty. Husserl arguably refers to the reversal of the applicatory act by it, and its use is legitimate as long as it does not obscure the *primary* abstractive base of the collective connection.

Corroboration for this "didactic" interpretation is found in a similar passage.[27] Husserl reiterates in it that his undertaking has been the "disclosing" of the "origin" of the concepts of multiplicity and number. "To this end," he writes, "it was necessary to keep in view precisely the concrete phenomena" from which they are abstracted. Such phenomena are the concrete totalities or "sets." However, cautions Husserl, "particular difficulties appear to block the way of the transition from these to the general concepts." This passage provides further indication that perhaps the abstraction of concepts from concrete totalities is not the *ordo essendi*.

In yet another passage, Husserl reiterates that "each concrete multiplicity falls under one, and indeed a determinate one of these concepts, to which 'is due a certain number.'"[28] Arguably, the term "falling under" functions as shorthand for the process of "attaching." Therefore, "it is easy," Husserl continues, "to characterize the abstraction which must be exercised upon a concrete totality before us, in order to attain the number concept under which it falls." This abstraction may well be interpreted as a regressive maneuver, similar to that synthetic one traditionally required of the analyst. Husserl describes this as a process in which the particularities of the entities collected are ignored and each is regarded simply as "something" or "one." The collective connection is finally seen and the determinate number concept of the particular multiplicity is attained.

Aside from the fact that this passage also seems to provide general internal support for the didactic thesis, descriptions of the abstracting of numerical concepts from acts are found immediately preceding and succeeding it. Unless Husserl was so confused as to juxtapose radically inconsistent positions in such proximity to one another, another interpretation must be sought of passages which seem *prima facie* to suggest that these concepts are abstracted immediately from concrete contents. An alternative interpretation is suggested by a close reading of the passages themselves, and it is in perfect concert with the central and coherent thesis of *BZ*. It goes without saying that the argument for this interpretation is not an argument for the ultimate correctness of Husserl's view as interpreted. It is to argue only that he maintained, and did not compromise, a radical separation of psychic acts and their contents, and that he was rigorously consistent in the conclusions he drew on the basis of this fundamental premise.

The view of Husserl's theory of the abstraction of numerical concepts combatted maintains specifically that the collective connection and, by implication, the "whole" for which it is responsible, is in some sense present amidst the contents. The passages often cited as evidence for this view must now be scrutinized to see if they provide the testimony alleged.

The Preeminence of the Totalizing Act: Refutation of a Prevalent Interpretation

It is true that *BZ* and most of Husserl's other pre-transcendental works adumbrate and motivate the later transcendental philosophy. However, the alleged "contentualization" of the totalizing act, the collective connection, or of the wholes to which each is tantamount, provides neither this adumbration nor motivation.[29] It does not, simply because the evidence for this allegation is hardly as compelling as some interpreters have maintained. All of these alleged "proto-noematic objects" boil down to acts, acts reflected upon, and concepts of acts abstracted by acts.

The question of the contentualization of the totalizing act or collective connection is hardly separate in *BZ* from the same question as it pertains to their respective wholes. Husserl insists that every whole, regardless of kind, is explicable by its particular connection. Clearly, in order for any "psychical whole" to be in some sense a content, its collective connection must *also* be a content relation. But it is precisely this which Husserl explicitly denies.[30] Even if it is held that the collective connection or its totality is perceptible with some more subtle intuition, it must still be explained why Husserl resorts to his reflexively convoluted psychic contortions when that which he seeks lies straightforwardly before him.

The following are those passages frequently cited as evidence for the contentual status of either the collective connection, its totality, or both.

(1) Husserl states here that it is a misunderstanding to say that the concrete totality consists "simply" of the individual contents. While he concedes that it is easy to overlook, there is, nevertheless, "still something there beyond the individual contents." This is "necessarily present" in all cases of totalities and can, therefore, be "noticed." This "something 'there,' " he writes, is none other than "the *connection* of the individual elements to the whole."[31]

While all that this passage says is true, it does not witness explicitly to a contentual connection or whole. It does not clarify where the connection is "necessarily present" and "noticed." This passage does not do the work even out of context that those who cite it would have it do.

While Husserl does not spell this out, it is likely that he assumes that the concept of collective connection, like that of something, was attached in an external and non-literal way to concrete collections. It may be that both of these constituent concepts of determinate numerical concepts are implicated in the attachment of each numerical concept to the concrete collection determined by it. It is incredible that Husserl should maintain that each enumeration involves actual recourse to the process abstractive of the number concept employed by it. It seems much more likely that each enumeration is an attachment of the given concept. It is Husserl's concern in *BZ* to indicate and clarify the primordial source of such concepts in the totalizing act and collective connection. The psychological elucidation of this source does not alter the fact that number concepts are experienced day to day in association with concrete groups. However this fact lends credence to the assumption that they are not only "contents," but even have their ontic roots in sensuous particulars.

(2) In another passage not long after that just considered, Husserl states that "the representation of the totality of given objects is a *unity* (*Einheit*) in which the representations of the indivdual objects are contained as partial representations." When compared to the manner of combination of the parts of other wholes, that of this whole is "loose and external." This is so much the case that "one would like to hesitate to speak here of any connection at all." But, cautions Husserl, "there is a peculiar unification (*Einigung*) there, and it would have to have been noticed as such because, otherwise, the concepts of totality and of multiplicity could never have been formed." The concept of multiplicity, Husserl continues, emerged in a manner analogous to that of the concept of other wholes, i.e., "through reflection on the manner of connection peculiar to them." The reflection in this instance, is on "the peculiar and, in its peculiarity, quite noticeable, manner of unification of contents, as every concrete totality features it."

Arguably, the "unity" mentioned here is none other than that of the "*psychical* whole" or unity of "parts" (although Husserl, at another place, cautions against this metaphor simply *because* of its analogy with other wholes and parts).[32]

Here also, as in (1), Husserl does not clarify where the manner of connection or unification is to be "noticed," where it is "featured." His manner of expression in this passage does seem to imply that the totality is something in and of itself, separate from or beyond the contents, and that as such it "features" the connection. At best, however, this is ambiguous and it can neither be concluded nor inferred that the collective connection is so featured *qua* contentual.

If no case can then be made for the contentual status of the collective connection, none can be made, by implication, for that of the totality. The central and consistent thesis of *BZ* is that this whole is tantamount to its contents being "thought-together." It is for this reason that Husserl endorses the view of John Stuart Mill that the "objects . . . form part of a psychical whole with reference to the 'thinking-together' act."[33] It is clear that the whole is "psychical" here *because* of the psychical act. In fact, it is the embrace of this act which *is* the whole. The latter is not a content because the act is not a content. It is not then surprising to find Husserl explicitly stating that "where a totality is given, our apprehension is, above all, purely preoccupied with the *absolute* contents."[34] There are, nevertheless, passages which might be taken to imply that this psychical whole is contentual.

(3) Toward the conclusion of the passage concerned with the whole of the rose, Husserl asserts that, in the case of physical connections, a "unification" is noticeable in the contents of the repesentation, while such is not discernible in those of collective connections. The latter therefore lack "intuitable unity," although (in the case of the totality) a "unity is present (*vorhanden*) in it and perceivable (*wahrnehmbar*) with evidence."[35]

It seems that Husserl is using "unification" (*Einigung*) and "unity" (*Einheit*) interchangeably, although he may be shifting in his use of "unity" to an emphasis on the whole. In any case, the last sentence, in which he employs the word "unity" twice, makes quite clear that the "perceivability" of such "present" in the totality is not "intuitable" though it is "evidential."

This could be quite true, once again, of the psychical whole. The "unity" of the latter is simply that provided by the thought embracing a number of otherwise disparate entities and aware of itself as doing so. *Their* togetherness or proximity is perceptible whether it is spatial or not. The *reason* for their proximity, *that* which gathers them, is not intuitable because it is an act and cannot, simultaneously, be a content.

Passages in which Husserl asserts that the "concrete totality as a whole" is the base for abstraction, rather than its "individual contents,"[36] may well be understood as referring to *psychical* wholes. The latter simply recur to the act as the basis for abstraction.

In another passage, Husserl asserts that the "representation of totality" is the "content" of the elementary act.[37] In and of itself this statement implies that "totality" is a content or is to be found among contents. Given the pronounced thrust of *BZ* in precisely the opposite direction of this implication, this is arguably Husserlian shorthand for "totality *of contents*." This interpretation is borne out by the passage stating that when a totality

The Concept of the Totalizing Act as Collelctive Connection

is given, it is the "absolute contents" that are apprehended, and not contents beyond them. Such shorthand, or even terminological imprecision, is not sufficiently unambiguous or ubiquitous in *BZ* to warrant erecting an alternative position on its basis.

Husserl's totalizing act is the very premise and foundation of the conceptual edifice he intends to clarify in *BZ*. And it is not only that this act provides the abstractive base, *qua* collective connection, for the concept of the latter. It, or an active psyche functioning very much like it, is also the abstractive means by which this concept is generated.

Husserl rejected the assumptions motivating his attempted "arithmeticization" of analysis in *BZ* almost as soon as it was written.[38] His basic theory of the generation of the number concepts of arithmetic remained in force some time longer (the arguments of this and the preceding chapter pertain as much to his *Philosophy of Arithemtic* of 1891 as to *BZ* of 1887). What is most characteristic of number concepts, and the concepts constituting them, is their ultimate derivation from *psychical acts*.

Notes

1. *BZ*, 334, 27—31/77, 8—10. The *PA* passage does not replicate the wording of that in *BZ*. The respective sections of each, referred to at this point by each (*BZ* 299, 3—302, 4/*PA* 17, 28—21, 11) are (except for a few minor alterations) identical.
2. *Ibid.*, 298, 21—24/16, 11—14. The point is essentially the same in the two texts. If anything, the emphasis in *PA* is placed more on the abstracting act than on the abstractive bases. The act relates itself to the concrete. See 301, 34—35/31, 2—3; 302, 7—8/22, 5—6; 309, 15—19/31, 15—21. In the latter passage Husserl distinguishes between "psychological description" (*Beschreibung*) of a phenomenon, and the "statement of its meaning" (*Bedeutung*), and between the objects of the description and statement, i.e., the "phenomenon as such" and that for which it "serves" (*dienen*), which it "signifies" (*bedeuten*). The phenomenon is the "foundation (*Grundlage*) for the meaning." As such, it is distinct from the latter.
3. *Ibid.*, 301, 38/21, 6. It is the concept of multiplicity which is referred to here as the general representation. The difference between these representations is discussed below in the second section of Chapter 3.
4. *Ibid.*, 299, 3—4/18, 5—6.
5. *Ibid.*, 324, 29—34/56, 17—22. Husserl observes that "difference" (*Verschiedenheit*) is not a content relation — as similarity is — but rather a "negative judgment" concerning the presence (absence) of such.
6. *Ibid.*, 309, 1—5/30, 24—31, 3. Husserl states that it is this which enters into the corresponding "general concept."
7. *Ibid.*, 300, 17—25/19, 22—30. What are so present here are the "continuous connections" which define the continuum — regardless of how different the "absolute contents" are case to case. The "concept of the continuum as a whole" arises (*entstehen*) through reflection on this connection.
8. *Ibid.*, 299, 5—13/18, 3—19. Husserl writes that it is those "characteristics" which are "common" that constitute the general concept.

44   Chapter II

9. *Ibid.*, 300, 9/19, 14. "Attending" (*achten*) is employed here as essentially synonymous with the "comparing" of different cases which exhibit the same "continuous connection."
10. *Ibid.*, 299, 5—13/18, 3—19. Concepts "arise" (*entstehen*) through this comparison of "specific representations," also referred to as "individual contents."
11. *Ibid.*, 314, 17—18/37, 5—6. The *PA* passage omits, "noticed for itself." Here it is the spatial order and position of entities which are not noticed and, consequently, do not enter into the concept. Such are "differences" between entities.
12. *Ibid.*, 337, 9—18. The passage is scattered to the far corners of *PA*. See *PA* 82, 8; 79, 1—27, especially 18—19. Husserl writes that, in the process of abstracting, there is disregard for the "particular determination" (*Beschaffenheit*) of the individual objects. "No special interest" is taken in the "peculiarities (*Besonderheiten*) of contents." Nevertheless, in the course of such (dis)regarding, the "concrete objects" do not disappear from consciousness. See also *BZ* 309, 25— 37/*PA* 31, 26—32, 2, where Husserl links such inattention to distinctions (*Unterschieden*) to our "intention" to unite contents. cf., *BZ* 314, 7—9/*PA* 36, 32—33 (lines 8—9 of *BZ* are omitted in *PA*).
13. *BZ* 300, 32/*PA* 20, 1. The concept formed in this case is that of a "whole." See 299, 5—13/18, 3—19.
14. *Ibid.*, 300, 17—25/19, 22—30.
15. *Ibid.*, 337, 24—34. The text of *PA* diverges from that of *BZ* before this point (at *BZ* 337, 9, and *PA* 82, 8). In another passage not reproduced in *PA*, i.e., *BZ* 334, 31—335, 1 (The immediate context of the passage corresponds roughly to *PA* 77, 4—27. *PA* 77, 28—80, 6, is essentially an interpolated expansion which rejoins *BZ* at 335, 34. There are allusions in this expansion to *BZ* 335, 14—335, 34, as well as an intersection with *BZ* 337, 13—18 at *PA* 79, 18—27. Generally this is new material.), Husserl describes the abstraction of concepts of wholes as linked to that of the concepts of their particular modes of connection. As a point of translation, it does not seem that a developed theory of the intuition of essences is implied here, as Willard's rendering of "*Beschaffenheit*" (*BZ* 334, 34) as "essence" suggests (see his translation of *BZ* in *Husserl*: *Shorter Works* [Notre Dame: University of Notre Dame Press, 1981], p. 115; hereafter *HSW*).
16. *BZ* 335, 1—7/*PA* 77, 10—15. The *PA* text omits "universal," as a prefix to "concept," with reference to "multiplicity."
17. *Ibid.*, 336, 2—13/80, 15—26.
18. *Ibid.*, 336, 21—38/81, 7—30. The *PA* version is more or less of an expanded paraphrase of its antecedent in *BZ*. While Husserl emphatically dissociates himself from the view that numbers are deposits in space-time, he observes that the lower number *names*, e.g., "*zwei, drei, vier,* etc. belong to the earliest creations of all languages." In this general regard he refers at *PA* 83, 25—35 to the "children" and the "wild peoples" and, in a footnote, cites some of the historical and anthropological works on which his views were apparently based.
19. By contrast, the contents of the totality are described as "absolute contents" (*BZ* 323, 15/*PA* 55, 1).
20. Ibid., 317, 7—17/45, 27—46, 7. The *PA* rendition of the passage makes essentially the same and significant point, although Husserl injects a critique of Baumann into it, thus expanding it relative to its antecedent in *BZ*. cf., *PA* 42, 23—43, 11.

   de Boer finds an inconsistency between the two versions which does not seem as apparent as he implies (*DHT*, p. 33). Husserl states in *BZ* that "numbers are mental creations (*Schöpfungen*) insofar as they form the results of activities which we exercise upon the concrete contents". In *PA*, he writes that "numbers are pure mental creations (the word "pure" is also used in the sentence in *BZ* preceding that just cited, and de Boer is correct in noting that Husserl says there that the "designation of numbers as pure mental creations of inner intuition involves an exaggeration and distortion of the true state of affairs") in a certain manner, and this is actually correct in that the numbers rest on psychical activities which we exercise on the contents."

Arguably, the phrase, "in a certain manner," is a Husserlian caveat here which de Boer fails to emphasize sufficiently, and that it is analogous in this regard to "insofar" in the *BZ* passage. If so, it renders the *PA* version essentially the same in meaning to its *BZ* antecedent. Husserl proceeds in *PA* to emphasize as well that while, given these caveats, numbers may be regarded as "mental creations," they are not entities in space-time. The emphasis of this in *BZ* in the statement that these are relational concepts, capable of being generated again and again, is omitted in *PA*. Perhaps this is an indication of Husserl's increasing uneasiness about identifying numbers with concepts of psychical acts.

One ought also consult the entire preceding Section IV of *BZ*, of which this passage is part, especially *BZ* 313, 1—34/*PA* 35, 12—36, 17, and *BZ* 315, 38—316, 26/*PA* 44, 6—45, 9. The first passage presents essentially the same arguments in *PA*, although the wording is altered and the text expanded at points. In denying that number is in any way a "part-phenomenon" in *BZ*, the term becomes "sensible quality" in *PA* (*BZ* 313, 8/*PA* 35, 18—19). In his critique of this general position of Mill and Lange, Husserl speaks of their "defiance" of the "clear evidence (*Zeugnis*) of inner experience" (*Erfahrung*; *PA* 36, 3—4). In the second passage, the two texts are essentially the same in point; Husserl injects more quotations from Baumann in *PA*. While *PA* retains the assertion that Baumann's theory rests on an "erroneous interpretation of the abstraction process" (*BZ* 316, 21—22/*PA* 45, 6—7), it omits the next sentence of *BZ* (316, 22—26): "Neither are they (number concepts) 'pure mental' creations of an 'inner intuition,' nor can one speak of a finding of the same in the external world, and of a being-together with and in space." However from what is retained immediately following the passage in *BZ*, it is clear that this omission signals no change in the theory (as is evident in the discussion of the first passage cited in the first half of this footnote).

21. *BZ* 316, 27—33/*PA* 45, 9—14; *BZ* 298, 10—18/ Roughly, the *PA* correlate to *BZ* 298, 7—20, is *PA* 14, 16—16, 11. This is an expansion of the passage in *BZ*, although the point of the latter passage is not changed. cf., *PA* 15, 21—24. In the introduction to *BZ* Husserl uses the term "*Verwendung*" (*BZ* 290, 30—38) when speaking of the "application" of mathematical concepts in (the context suggests) the sciences and their attendant technologies. "To apply" is undoubtedly a more elegant translation of "*knüpfen*" than "to attach." The latter rendering is given because it conveys "externality" better in English than the former. Husserl does employ "*verwenden*" when speaking of language use, and perhaps the implication is that words seem to "clothe" their referents most smoothly.
22. *BZ* 338, 14—16.
23. *BZ* 336, 13—15/*PA* 80, 26—29. For a discussion of the negative judgment which asserts the non-identity of terms, and, while sometimes mistaken for a content relation, is made on the basis of terms but not found among them, see *BZ* 324, 10—38/*PA* 55, 35—56, 26.
24. *Ibid.*, 298, 14—18/14, 16—16, 11.
25. See "On the Concept of Number," *HSW*, p. 96.
26. *BZ* 298, 19—24/*PA* 16, 9—14; and 299, 3—7/18, 5—9. The first *PA* passage diverges in wording at points, but the central point is the same. Husserl omits in it the characterization of totalities of determinate objects as gathered "indiscriminately" (*beliebig*) in addition to "arbitrarily" (*willkürlich*). In the second *PA* passage, although the wording is somewhat different, the point of its antecedent is retained. In *PA* Husserl does not emphasize the "from which" abstraction is made in the context of a query (*BZ* 299, 7/*PA* 18, 9).
27. *Ibid.*, 327, 6—12/64, 6—12. The *PA* passage is virtually identical to that of *BZ* with

the exception that Husserl replaces "set" (*Menge*; *BZ* 327, 10) with "multiplicity" (*PA* 64, 9—10).
28. *Ibid.*, 336, 39—337, 18/81, 33—82, 8. The *PA* passage essentially repeats, with minor changes, the *BZ* text up until *BZ* 337, 9. Some of *BZ* 337, 9—18 is found, or essentially the same point is made, in *PA* 79, 6—24. As noted above, *PA* diverges from *BZ* after *BZ* 337, 9, and at *PA* 82, 9.
29. Admittedly a barbarism, the term "contentualization" was suggested by Dorion Cairns. In his *Guide for Translating Husserl* (The Hague: Martinus Nijhoff, Phaenomenologica, vol. 55, 1973), on p. 74, he writes that "*inhaltlich*" may be translated "contentual." It is hoped that its awkwardness will not subvert its indication of what might also be termed "content-status."
30. Husserl appeals to the high court of "inner experience" for validation of his position on this point (*BZ* 333, 18—23/*PA* 73, 25—30). He emphasizes the point in *PA* by adding that this is a result which "repeatedly intrudes" in the critical discussions of the previous chapters. With regard to "inner experience" see *BZ* 304, 23—28/*PA* 24, 29—34, as well.
31. *BZ* 299, 26—32/*PA* 18, 31—37.
32. *Ibid.*, 335, 7—10/77, 16—21.
33. *Ibid.*, 332, n. 1/73, n. 1. Also, as Husserl writes at *BZ* 317, 4/*PA* 45, 23, the "totality" of collected objects is simply convertible with the fact that "they are one" in virtue of the act.
34. *Ibid.*, 323, 14—18/54, 39—55, 4.
35. *Ibid.*, 332, 6—16/72, 23—29. As already noted, *PA* 72, 23—29 recapitulates, almost verbatim, *BZ* 332, 6—12. However *PA* omits the passage cited here from *BZ* 332, 12—16. See Appendix II.
36. *BZ* 299, 17—19/*PA* 18, 23—25.
37. *BZ* 335, 1—7/*PA* 77, 10—15.
38. See, e.g., his letter to Carl Stumpf (c. 1890) reprinted in Edmund Husserl, *Studien zur Arithmetik und Geometrie. Texte aus dem Nachlass* (1886—1901), edited by Ingeborg Strohmeyer, *Husserliana* XXI (The Hague, Boston, Lancaster: Martinus Nijhoff Publishers, 1983), P. 245.

Willard maintains that Husserl's departure from the prevailing assumption that analysis could be arithmeticized, was due to the fortuitous conjunction of his writing the last part of *PA* dealing with this, and his examination of Schröder's *Vorlesungen über die Algebra der Logik* of 1890, in preparation for his review of 1891 of this work. It seems to have been in this general period that Husserl became persuaded that universal arithmetic was not *derived* from numerical concepts after the manner of symbolic numerical concepts, because it was "a piece of formal logic" and hence, to be *applied* to numbers. As a "symbolic mechanism," it "can be assigned different interpretations" and can "thereby, be applied to various distinct (though analogous) conceptual domains, not to that of number alone" (*LOK* 109). Consequently, insofar as this formal arithmetic, its *ad hoc* concepts, and the formal logic of which it was part, could no longer be regarded as derivative from, and founded upon, actual number concepts, they could not be regarded as founded at all. His attempt to do so in *BZ* and *PA* is, in Willard's famous characterization of it, "the logic that failed." It was its failure that turned Husserl in the direction of the research that culminated in the publication of the *Logical Investigations*.

CHAPTER III

Symbolizing: Prosthesis of the Totalizing Act

Up to this point the totalizing psyche is preeminent and apparently omnipotent. Yet a challenge to this preeminence is sounded in *BZ* itself when Husserl states that all objects agree only in that they are "contents of representations," or are "represented by means of contents of representations in our consciousness." There are entities which are not immediate contents of representing consciousness, which are not "in" it as such.

But this challenge is simultaneously its own silencing. Husserl demonstrates in his *Philosophy of Arithmetic* (hereafter *PA*) that while the totalizing psyche is limited and finite, its finitude is only the occasion for it to demonstrate its truly regal powers. Like all rulers whose totalizing aspirations outstrip their actual grasp, it extends itself to far-flung dominions through proxies which represent it. These function as "prostheses" of the finite psyche enabling its grasp to extend infinitely.

As has been demonstrated in the preceding chapters, the totalizing act is at the center of both *BZ* and *PA*. It alone is the totality. From itself it generates concepts founding number concepts, all of which refer and defer to it as their progenitor. Like a sovereign, the psyche is the source of the unity of its subjects, but never mingles with *hoi polloi*. It not only maintains a regal distance, but is incapable of being touched by that altogether other and different from it in kind.

The totalizing act remains at the forefront of *PA* proper. Indeed, it has multiplied and presents itself as a complicated complex of totality upon totality. These remain no more and no less than acts upon acts.

Husserl also elaborates the structure or anatomy of the abstracted concept. It becomes clear that there is a lacuna of sorts between this *abstractum* and its abstractive base in psychical acts. This gap will eventually widen to the point that the origin of the concept in the psyche will be rejected altogether.

It is within the context of his discussion of concepts that Husserl treats those of arithmetic and distinguishes between those that are "actual"

48   Chapter III

(*eigentlich*) and those that are not. "Inactual" number concepts witness simultaneously to the psyche's finitude and to its capacity to grasp symbolically that which it cannot grasp "actually." Its capacity for symbolizing functions as but a mode of the totalizing intent of the psyche.

Husserl's invocation of the complementary totalizing-signifying modes of the psyche for the conquest of pre-given sensible multitudes will be presented in Chapter IV. It becomes increasingly clear that he is not in the least averse to deconstructing and reconstituting sensibilia (even sensible particulars) in the image of the totalizing psyche. Husserl is quite frank in acknowledging that it is precisely this that he is doing. Pre-given wholes have an integrity *only insofar* as they have been reconstituted and assimilated by the totalizing psyche. This, he writes, is its "primordial intent" which it also enacts in the sensible realm by means of proxies offered up by sensibilia itself.

The Hierarchic Complication of Totalizing Acts

Husserl insists that his primary intent in *PA* is not architectonic but fundamental. It is not to build a "closed system of a philosophy of arithmetic" but to lay "foundations" which have been "secured."[1] And *psychical acts* continue to be the most fundamental of foundations in *PA*.

In the course of arguing that disparate entities are related only in being thought-together in one act, Husserl injects an expansion into the original text of *BZ*.[2] In this interpolation he argues that regardless of the number of members of the collections, "the colligatory act surrounds all the members without separate collective connections." Regardless of the number of *colligata*, the act is one and is simple. It is not a "complex or web of collective connections."

Husserl builds on his analysis of the totality in *BZ/PA*. He describes the ascent beginning with disparate entities and their initial totalization, and continuing upward through successive totalities, each comprehending its predecessors. Husserl not only clarifies the relations "more" and "less" as they pertain to the series of concepts of determinate totalities. In doing so he reiterates that these "totalities" of increasing complication are no less psychical acts (upon acts) than they were in *BZ/PA*.

Before even the smallest totality can be generated, each of its contents must be apprehended by a "particular psychical act." The "grasping-together" of these contents requires a "new act" which is a psychical act of the "second order." If this second order act has collected two contents, and if two other such acts have done likewise, we may then, writes Husserl, form a "multiplicity of six objects" mediated by the "sub-groups" and their correlative acts. This act of collecting or adding which forms the "embracing collective unity" is then classified as a psychical act of the "third order," and so on. Husserl characterizes this hierarchy as a "complication of acts directed upon one another." If a totality is tantamount to its contents being thought-together, then larger totalities comprising this one as well as others are no less what they are in virtue of a totalizing act which "embraces all members."

The complications of totalities and their correlative "psychical acts of a higher order" are treated more explicitly.[3] In the context of comparing Totality [A,B,C], and Totality [A,B,C,D,E] (the judgment that the second is "more" requires the simultaneous representation of both sets and the recognition that the first is a subset of the second), Husserl explicitly links the "capability (*Fähigkeit*) that we have" with the complication of totalities described. For just as the psyche is capable of totalizing several entities, it is also capable of totalizing several "totalities . . . into one totality" without "losing their particular connections." We are capable of representing totalities, Husserl observes, "the elements of which are also totalities." Even "totalities of totalities of totalities" are thinkable.

Husserl states explicitly that the "psychological foundations" of these "complicated formations" are "higher order psychical acts" which are "again directed upon psychical acts." This entire hierarchy of increasingly complicated tiers is founded on what he terms "simple totalities." These are simple because their constituents are not totalities but "not further analyzed individual contents." These contents are those totalized by second order acts and residing in their embrace as "simple totalities."

The ultimate foundation of this inverted pyramid is not even simple totalities, but the "first order act" which apprehends each of the contents in and of itself. It is this first order act, writes Husserl, in which the "particular connections of the 'part-totalities' rest."

Husserl maintains that when a totality is given, it is the "absolute contents" which are primarily apprehended.[4] Yet these "primary contents" are not apprehended directly, but are mediated to the totalizing act by the representing acts of the "first order."[5] "Second order" acts are not directed upon primary contents immediately, but upon first order representing acts. In *PA*, Husserl clearly separates the apprehending of absolute contents from the totalizing of second order acts. If this apprehending was not performed by the "noticing" facet of the totalizing act in *BZ*, then neither was it clearly distinguished from this facet. Husserl's characterization of the contents as "fundaments" of the totalizing act in *BZ* may be an allusion to this distinction. In any event, whether the contents of the second order act are absolute contents for it immediately, or only mediately, the simple totality is itself *never* an absolute or primary content. The totalizing *act* remains preeminent. Yet it is *not* the most fundamental of acts, it is only "second-order" — for the time being.

Even as it may be an object of reflection in which its collective connection is discerned (or the concept of such abstracted), the simple totality may also be an object of reflection for the third order act uniting it with other simple sub-groups. Husserl insists nevertheless that the "immediate content" of either reflective act is the "relation-endowing (*Beziehungstiften*) act." Only "by means of" this act are primary acts and their absolute contents mediated to higher order acts.[6] The totalizing act remains the context for understanding number concepts in *PA* proper.

The Anatomy of *Abstracta*

Husserl's theory of the abstraction of the concepts constitutive of number concepts is the same in *PA* as in *BZ*. His description of "totalities of totalities of totalities" as boiling down to acts upon acts provides further evidence that the abstractive bases of number concepts are ultimately acts, and that such concepts are ultimately *concepts of* acts. *PA* also provides further clarification of the structure of the concept in general and its relation to its abstractive bases once it is abstracted. Number concepts, and their division by Husserl into those "actual" and "inactual," are best approached from within this broader context.

In an interpolation of the *BZ* text there is an early description of the abstracted concept's structure as that of increasingly differentiated emanations from an austerely unitary apex.[7] Husserl cites the example of the concepts of multiplicity and collective connection. As in the concrete, so, he argues, in the abstract the concept of multiplicity reduces to that of the collective connection.

The term "concept" may refer to the undifferentiated apex of the concept of *abstractum*. In the sense of logic, Husserl writes, the *abstractum* collective connection is the "meaning" (*Bedeutung*) of the name "multiplicity." The distinct concept multiplicity thus resolves itself into the *abstractum* of the "collective connection as such." Yet this resolution is not entirely satisfactory, Husserl points out, because reference is not made only to pure meanings when employing the term multiplicity. Reference is made as well to the "collective whole." This reference, he insists, is to "something" which "possesses" (*besitzen*) this abstract meaning. This is the "object of interest" which is also conceptual. Husserl speaks here of "applying" (*verwenden*) the name "multiplicity," and it is in so doing that reference to the collective whole emerges.

The structure of the concept as simultaneously unified and differentiated is clarified by Husserl through further examples. In the same passage he cites that of the expression "the concept 'Man'" which is "equivocal." The expression may refer to either the "universal" (*allgemein*) or to the "abstract concept." The former is the concept of "man pure and simple," of something that possesses certain abstract "characteristics" (*Merkmale*). The linguistic formula Husserl employs here is essentially the same as in the previous example when he wrote that "something possesses this abstract moment of the collective connection" (by inference, this something was the universal concept).

At another point in *PA* Husserl clarifies this lower or universal story of the concept in relation to the concrete.[8] He states that the "universal presentation, 'a man'" cannot serve as a "surrogate" for the "determinate man," Peter. In order for it to be an adequate surrogate for him, characteristics peculiar to Peter must be "added." It is clear that this process of conceptual approximation to a particular is accomplished only through a complication of *concepts*. A configuration of such may be constructed which adequately indicates the determinate entity. Nonetheless, this approximation to particularity is not effected through any further descent through or from conceptuality. The characteristics constituting the configuration indicating Peter remain universal concepts.

A "particular interest" may be taken neither in the determinate nor indeterminate possessors of the characteristic, but in the "characteristic for itself." Interest may be taken, further, in an entire "association" (*Verein*) of such, and this association of concepts "for themselves" is the *abstractum*. In this instance it is the "*abstractum*, 'Man.'"

Husserl does not maintain that the interest taken in one facet or another of the concept is somehow generative of that facet. The concept *in toto* has already been generated by the abstractive act in which attention does play a major role. The interest of which Husserl speaks is revelatory of whatever facet of the concept to which it is directed. Husserl refers to the entertaining of the *abstractum* as involving a "particular" or "special" interest. This suggests that such is not the usual direction of attention. This is corroborated in another passage contrasting this particular interest with the ordinary interest in universal concepts motivated by what Husserl terms "practical necessities."

In this passage he describes the abstractive ascent from concrete particulars to concepts, the heady descent from the homogeneity of *abstracta* on waves of differentiation, and the intimate interpenetration of language at every point.[9] Husserl begins by noting the "double meaning" of every abstract name. The name may refer to the "abstract concept as such" (undoubtedly in virtue of the particular interest in the latter). However, interest is ordinarily taken in "concrete things and relations." Language is mainly employed by interests motivated by practical necessities. While language is incorrigibly conceptual, fraught with abstract names, these may be used to designate "concrete things and circumstances." Several general names connected with one another "single-out" and articulate the concrete particular.

Husserl regards language as most fitted for contemplation of concepts. Language-users are nevertheless subject to practical concerns in which interest is necessarily taken. Consequently, language is used as a tool facilitating practical dealings with concrete things and circumstances, rather than as a map to the region of pure meanings. To "employ" (*verwenden*) it toward practical ends is simultaneously to "apply" it. Husserl characterizes both psychical acts and the concepts abstracted from them as "attached" or "applied" to concrete phenomena. This employment of abstract names amounts to their compulsory descent to universal names, a descent corresponding to that of the concept. Concrete particulars are singled-out by both universals, or perhaps by the *articulated* concept. Concepts are applied to concrete phenomena in speech acts expressing interests necessitated by practical circumstances. Underlying Husserl's theory of concepts and their linguistic correlates is one of "interested" referential psychical acts.

Husserl makes the same points using the example of color. He speaks first of color which is the "logical part" common to red, blue, etc., and later speaks of the "abstract names" corresponding to these concepts.[10] Color, as exhibited by the instances cited, is the high and homogenous *abstractum*. However, when speaking of this color and that color, general names are employed for "individual species as such." These are presumably "Red," "Blue," etc. This "manner of application" (*Verwendung*) may be even more sharply determined in referring to "certain colors," e.g., Crimson, or Vermilion. The *abstractum* may be referred to as the "abstract concept, 'color' " or simply as the "concept color." But the plural, Husserl warns, cannot be used to refer other than to the "objects of the concept" (*Begriffsgegenstände*), colors perceived as similar.

This descent from *abstracta* does not precisely recapitulate the ascent from phenomena. Originally, Husserl writes, the "adjective 'red' " functions as a general name for all red things. From that adjectival function emerged the "abstract name 'Red.' " In the process of application the latter becomes not a general name for things but for the "differentiations" of the abstract concept "Red" — "Crimson," "Vermilion," etc.

While emphasizing the practical concerns and interests which determine the descent from *abstracta*, Husserl is quick to point out that the "motivation" (*Veranlassung*) to reflect on the "*abstracta* as such" is by no means absent from "life and science," and that language "provides" as well for this direction of attention. For instance, "hood" is added to "father," or "kind" to "human," so as to facilitate the designating of "abstract characteristics" (*Beschaffenheiten*) once interest is taken in them for themselves. This interest in or reflection on the *abstractum* is *subsequent* to the original interest in it which was "only . . . insofar as this object and that object contained it as a characteristic." Husserl states explicitly that the abstract concept was already formed when the general name emerged but was an object of interest only insofar as it was exhibited by given objects as a characteristic. The interest in the "*abstractum* for itself" emerges later.

According to Husserl, while language is essentially conceptual, it is but the articulation of more primordial concept formation. It is borne by the direction of interest and is adjusted by the latter to the more coarse or subtle demands of the respective objects of interest. Practical concerns and jargon therefore presume the *abstracta* as their conditions at all points. *Praxis* may become aware of its absolute indebtedness to these conditions when it finds occasion to catch its breath and reflect. At this point it may gradually redirect its attention, contemplate, and even chart the conceptual cosmos. This redirecting of attention facilitates its recognition of concepts as *abstracta* produced by its *own* abstracting interests.[11]

Husserl's discussion of the conceptual stratum of arithmetic is carried on in essentially the same vein. He observes that the "determinate number (concepts)" of, e.g., "*zwei, drei, vier*," etc., exhibit a similarity which "we indicate" by means of the universal concept called by its universal name of "number."[12] Conversely, this concept is "accounted for" (*erklären*) by reference to this similarity. Husserl emphasizes that the universal concept is not a "physical or . . . only metaphysical" part of these concepts. The relation between the universal concept number and the specific number concepts seems to be that between "logical part and logical whole." Husserl notes as well that (contrary to the nominalistic doctrine of the elder Mill) the universal name does not simply indicate the similarity of specific names but also that of the concepts to which such names refer.

These points are amplified and clarified when he argues that it is emphatically not the number concepts which are subjects of various arithmetical operations.[13] These "number concepts as such," or *abstracta*, are "self-identical." Because each concept remains always what it is, it is impossible to add, e.g., the concept "Two" and the concept "Three" and to derive from them the concept "Five." *If* one were to attempt to add these concepts (2, 3, and 5) one would not derive 10 but 3 (just as in the case of adding one apple to one apple to one apple). According to Husserl, it makes no difference whether one is adding concepts or apples.[14] He insists upon the fundamental unity of *abstracta*, not merely of the concepts of numbers but of all concepts. They are self-identical, indivisible, and incorrigible.

The arithmetician, he writes, does not normally operate with such *abstracta* but rather with the general objects of these concepts. The sign, "5," e.g., does not then "signify" (*bedeuten*) the "*abstractum*, Five." It is, rather, a universal name referring to "any arbitrary set as such falling under the concept 'Five.' " This referent is again simply the concept of something which exhibits the given characteristic (in this case, the determination of elements called "Five").

These number concepts are divided by Husserl into those that are "actual" and those that are not. These concepts turn out to provide further insight into the capacities of the totalizing psyche.

THE SELF-EXTENSION OF THE TOTALIZING ACT BY PROXY

In his discussion of "totalities of totalities of totalities," Husserl makes it quite clear that each totality of this hierarchy remains tantamount to a totalizing act. This hierarchy of totalities is one of higher directed upon lower acts.

As concepts of totalities, the various number concepts continue to be those of totalizing acts in *PA*. This is true whether they are generated in the serial manner of *BZ/PA*, or that of an inverted pyramid of sets comprehending sets of *PA* proper. In either instance the particular determination of entities is grasped by one totalizing act.

Husserl defines the totality in *BZ* as a group of entities each of which is "noticed 'for itself,' " and, simultaneously, held "together, unified, with the others" of the group embraced by the totalizing act. The totality is therefore defined not merely in terms of the embracing act, but in terms of this *capacity* of it. The totalizing function of the act is defined *as* this capacity to see each member of the totality for itself and all of the others, simultaneously, in the same way.

This is precisely the formula Husserl invokes in *PA* when he characterizes the actual representation of multiplicities. The psyche must be able to attend "actually (*faktisch*) . . . in one act, to each of" the "members as one noticed for itself, together with all the others."[15] But this capacity is quite limited. Husserl observes in *BZ* that a "*de facto* limit" is reached in the thinking of an indeterminate multiplicity. It might be that this limit is precisely that when the psyche can no longer maintain its grip on the multiplicity. Husserl states in *PA* that it is "only under particularly favorable circumstances" that the psyche can "actually represent concrete multiplicities of approximately a dozen elements . . . twelve is . . . the extreme limit (*letzte Grenze*). . . ." He sets this "outside limit" at different points at different times. Regardless, the totalizing act is limited in how far it can extend itself.

Husserl concurs, for he writes that we are "highly limited in our representational activity." These limitations are tantamount to the "finitude (*Endlichkeit*) of human nature."[16] Specifying the locus of this finitude, he speaks of the "essential imperfection" of our intellects.

The problem Husserl faces is this. Number concepts are concepts of totalities, and the latter depend solely on the totalizing *capacity* of the psyche. Since this capacity does not extend beyond the grasp of totalities exceeding twelve members, arithmetic must then be limited to the first twelve cardinal numbers. These, as the concepts of actually represented totalities, are "actual" number concepts. Their characterization refers directly to the capacity (and lack thereof) of the act.

But the arithmetician traffics with much larger and infinitely more complex numbers which cannot be concepts of actual totalizing activities. "The presupposition," Husserl concludes, "that each arithmetical operation is an activity with and on the actual numbers (*wirklichen Zahlen*), cannot be in keeping with the truth." This presupposition is both "naïve" and one of "common practice." It is false because it fails to do "justice to the fundamental fact that all number presentations ... beyond the first few in the number series, are symbolic, and can only be symbolic." This, he adds, is "a fact which truly determines the character, sense, and purpose of arithmetic entirely. ..." Husserl proceeds to make the same point with regard to the attendant operations beyond the first few numbers.[17]

The greater part of arithmetic can only be symbolic. It is fraught with inactual concepts. This means that arithmetic owes its being and practice to human finitude, to the imperfection of the intellect, to its incapacity to totalize actually entities exceeding twelve in number. Husserl leaves no doubt about this, for he observes that *if* we had "actual representations of all numbers, as (we do) of the first in the number series, then there would be no arithmetic ... it would be completely superfluous." If we had actual representations immediately in one act, of "the complicated relations between numbers which, now, become revealed only troublesomely, through complicated calculations," then these entities would have for us the same "intuitable evidence" as does, e.g., the proposition "$2 + 3 = 5$" (or more precisely, as do the totalities designated by its signs). They would be clear, Husserl writes, "immediately and with evidence." Arithmetical calculation through which access to these inactual concepts is gained is, he writes, "nothing else than a technical means" enabling us to "overcome the essential imperfection of our intellects."[18]

On the one hand, arithmetic is a wholly contingent *technē* rooted in and motivated by the limitations of human intellect. Husserl approvingly cites Dedekind's reformulation of the dictum of Gauss. It is *anthropos*, not *theos* that is the arithmetician. Arithmetical practice backhandedly signifies fundamental human limitations. On the other hand, while it is "nothing else than a technical means," it is nevertheless that by which the psyche overcomes its limitations. The latter is not infinite in its actual capacities, it is capable of "capturing" only a small number of contents. Yet its reach exceeds its grasp, and it devises magnificent symbolic prostheses to grasp that which eludes its actual embrace. The psyche's symbolizing capacity waxes when its actual totalizing capacity wanes. The two are of a piece. *PA* may therefore be understood as an expanded meditation on one line in *BZ*: "Wherein all objects — actual and possible, real and not real, physical and psychical, etc. — agree, is only this, that they are contents of representations or become represented (*vertreten*) by means of contents of representations in our consciousness."[19] This formula of "inactual representation" is that of a broad psychical maneuver not limited to mathematical symbolism. The universal concept "Peter" is, e.g., an inactual representation of the determinate man. Husserl has founded the first few number concepts in actual totalities. Yet that leaves most of arithmetic unaccounted for. How, he asks, can concepts be employed "which one does not *actually* have?" Isn't it "absurd," he presses, to found the most certain of all sciences, arithmetic, on such concepts? His answer is that such concepts are had inactually by means of symbols. But *how* may something be "represented by means of contents of representations" when no actual encounter has ever been had with it?

There are groups the size of which prevents their grasp as instances of particular numerical determinations. They may be enumerated, given a sufficiently long series of number names. There are also groups, to use Husserl's examples, of a million or a trillion members of which, in the absence of a Cray supercomputer, one would "truly need the light years of the astronomers to gain the actual (*wirklich*) representation."[20] These multiplicities cannot be "run through" in several human lifetimes, let alone be actually grasped as totalities. Consequently, their actual number concepts are "inaccessible to us," Husserl writes.

Instead of having actual concepts of these groups, there is no alternative but to "operate with sharply determined symbolic surrogate concepts. . . ." These concepts, whether of numbers or of operations on numbers, "represent" (*vertreten*) a "determinate actual" number or operation concept — "although (such is) not actually practicable" (*ausführbar*).[21]

The concept of, e.g., one million, mediated by its sign or symbol, is that of a non-practicable totalization. It is a symbol of that which the psyche would do if it could were it not finite and limited. The signs derived by arithmetic calculations and operations signify what it cannot accomplish in fact, but which it accomplishes contrary to fact or symbolically. "The systematic numbers offer," Husserl writes in a great paean, "a uniform and (by idealizing abstraction from certain limits of our capacities) inexhaustible method of extension of the number sphere beyond every boundary."[22] Limited by the number twelve, the psyche abandons its attempt to maintain an actual grasp on groups, and rules the latter instead by the proxies of abstract names or signs. The latter, wholly within its actual grasp, suddenly extends its totalizing intentions without limit.

In another passage, Husserl asserts that "symbolic formations represent the inaccessible (*unzugänglich*) number concepts 'in themselves.'"[23] What is "*an sich*" is not a Kantian object of any sort, but totalizing activity. Since it is accomplishable at lower and less complicated levels, it is not inaccessible in principle. At the point at which it is not accomplishable and therefore inaccessible, the "systematic of number concepts" supplies "a systematic manner of formation," and this is a "symbolic representative" (*Vertretung*) for the "*absent actual* number concept" (of the absent totalizing act.)[24]

Husserl leaves no doubt that there corresponds to the formation of every system of signs, "in rigorous parallelism," a system of the formation of concepts. And each is founded, respectively, on "fundamental signs" and "fundamental concepts."[25] These fundamental concepts are those also termed "actual." As the concepts of actual totalizing acts, they are also symbolized. Fundamental signs may be employed without reference to their abstractive bases of determinate totalizing acts. And elementary operations may be performed on them which generate further actual or inactual signs. But, an even "more rigorous parallelism" exists between this "method of extension of the series of number concepts" and that of "the series of the signs of the numbers."[26]

This "parallelism" is characterized in the strongest possible terms by Husserl. The "systematic of numbers" (especially the decadic system) is not primarily concerned to "sign concepts which are given," but "to construct (*konstruieren*) new concepts and to sign them, simultaneously, with the construction."[27]

This passage tends to leave the impression that these concepts in some sense *precede* their signification. They are not, as actual concepts are, given and then signed because the arithmetician is already operating with signs, with symbols, with names. As Husserl writes, "in this pure external procedure" of calculation, "names are derived from names," and, "ultimately, names result whose conceptual meaning, as the result sought, is necessarily produced" (*ergehen*) by this procedure.[28]

It is not startling that Husserl should describe inactual number concepts as being produced. Actual number concepts are described similarly. The salient difference between the two modes of production is that the abstractive base is present in the latter and absent in the former. This is the difference expressed by their names.

The signs of the actual concepts may be related, through, e.g., the

Chapter III

operation of multiplication, to produce new names (which are very likely inactual). Husserl is vehemently anti-nominalist in insisting that there are concepts corresponding to these names. But these concepts are not abstracted from actual totalizing acts. They are intimated by their respective names. The latter then indicate concepts of unexecuted totalizing acts.

What are "constructed" via signs (which are also constructions) are concepts of non-existent activities. Somewhat paradoxically, the psyche symbolizes the concepts of activities it has not abstracted because it has not, or in most instances could not, perform the activities themselves. It then signifies that which may well be impossible for it to do. In so doing, it has, strangely, turned its deepest limitations on their heads. The psyche is not only the signifier in these calculations, but it is also the signified. It is eminently self-referential. Inactual number concepts are also concepts of (albeit non-existent) psychical acts.

Husserl himself ultimately came to reject the latter proposition — even in its more plausible pertinence to "actual representations" of multiplicities. He did not, however, reject symbolizing as a fundamental psychical extension of the totalizing embrace by proxy to that which it could not surround immediately and in fact.

Husserl's unremitting anti-nominalism moved him to insist without fail that there are conceptual correlates of "names derived from names." Nevertheless, it is not difficult to understand the ease with which the "abstraction" from the meaning of signs is accomplished, the ease with which the "definitions of number and rules of operation" are "supplanted" (*ersetzen*) by the "corresponding formulas of equivalence of connections of signs." Husserl defines arithmetic in terms of just such an "independent system of signs." As such, it makes possible an "enormous savings of psychic work," as well as an enormous expansion of intellectual productive power in general.[29] He is nevertheless deeply concerned that arithmetic not become merely an "empty play of signs,"[30] a practice with "wholly arbitrary, meaningless signs (like pieces on some gameboard)."[31]

Husserl emphatically distinguishes this "art of calculating" (*Rechenkunst*), from the "art of arithmetical knowledge" (*Erkenntnis*).[32] The latter is gained by elucidating the "external and blind processes" of this *technē*, this fetishism of signs by which even the "great mathematicians" have been lulled into somnambulance. This elucidating is simply the bringing to consciousness of the "conceptual content" of the "contentless signs." The knowledge of this content opens the eyes of those engaged in symbolic processes "blindly and habitually," those falling prey to the confusion of "signs and things."[33]

Husserl's endeavor to "arithmeticize" higher mathematical analysis, thereby giving a firm grounding to its auxiliary concepts, is of passing historical interest. Many mathematicians believed in it, but Husserl forsook it early on in the belief that it was fundamentally misguided. It would be a mistake to infer from this that nothing remained that would significantly influence his thought — even in mathematics and logic. The symbolical activity of the psyche elaborated in this particular context not only endured, but continued to flourish and gain momentum as a fundamental force in Husserl's thought. As such, *it is but a mode of totalizing activity.* When the latter wanes, it waxes magnificently.

Husserl demonstrates the intimate collaboration of symbolizing and actual totalizing in his account of the psyche's attempted totalization of sensible multitudes. The fact that these present themselves as pre-given unities does not daunt the imperious psyche. They must be deconstructed and reconstituted by it.

Husserl's doctrine of concepts is complete in *PA*. Yet the totalizing act employed originally in the founding of arithmetic is not abandoned by him. It, and its symbolic mode, continues to be the means by which he seeks to account for not only sensible multitudes, but their individual constituents.

NOTES

1. *PA* 287, 5—12.
2. *Ibid.*, pp. 74, 20—75, 16. In the original *BZ* form of the passage just cited, (*PA* 74, 14—19/*BZ* 333, 35—334, 3), the "*einem*" was emphasized by Husserl.
3. *Ibid.*, pp. 90—91.
4. *Ibid.*, pp. 54, 39—55, 4 (*BZ* 323, 14—18). See also *PA* pp. 42, 6—8, where Husserl indirectly corroborates this point. There he observes that the concept of the "connection" is not to be found in the contents, but only in the acts — and, as a result of this, it is found only in *reflection* on the acts.
5. A few lines earlier, Husserl uses the term, "mediation" with reference to "psychical acts." These are now specified as "acts of the first order."
6. *Ibid.*, pp. 69, 35—70, 1. See Appendix III.
7. The expansion is found between *PA* 77, 28, and 80, 6. The passage considered here is located at *PA* 78, 1—30.
8. *PA* 193, n. 1.
9. *Ibid.*, 136, 30—138, 26.
10. *Ibid.*, 137, 4—5; 138, 15. cf., also, 82, 26—9.
11. The similarity in structure of this and later accounts, where Husserl recurs to constituting acts on the basis of objects *qua* "transcendental clues," does not entail the conclusion Biemel believes it does (see discussion in Appendix II.)
12. *Ibid.*, 82, 9—29.
13. *Ibid.*, 181, 11—183, 8.
14. *Ibid.*, 182, 18—29. Husserl does not speak explicitly of "concepts" here, but rather of "numbers." It seems that he has number concepts in mind, however.
15. *PA* 192, 8—18.

16. *Ibid.*, 191, 24—35.
17. *Ibid.*, 190, 18—30.
18. *Ibid.*, 192, 2—30.
19. *BZ* 336, 6—10/*PA* 80, 19—23.
20. In a footnote on the same page, Husserl refers to concepts which we "do not actually (*eigentlich*) have." He uses these terms interchangeably. See Appendix IV.
21. *PA* 263, 7—18.
22. *Ibid.*, 260, 1—5.
23. *Ibid.*, 223, 33—36. See also *PA* 261, 16—38. de Boer's reading of *PA* in this matter of "number concepts 'in themselves'" seems correct. He cites Farber who (following Frege) understood these as extra-mental entities of the Kantian variety, and viewed Husserl's terminology as "semantic sloppiness," rather than indicative of a deeper confusion of "subjective/objective" spheres, as Frege believed it to be (*DHT*, pp. 27, 64, and 118). Willard (*LOK*, 111) also seems to understand the "number concepts 'in themselves'" in this fashion, and shares the belief as well that the root of the problem is Husserl's "fateful choice of terminology" (*HSW*, p. 88).

    This controversy is really but one *persona* of a much deeper divergence in interpretation of Husserl's presentation of number (specifically: whether it is fundamentally rooted in the totalizing act or in its contents — whether it has "contentual" status or not).
24. *PA* 237, 8—30.
25. *PA* 228, 30—7. A similar point is made at *PA* 259, 15—34.
26. *Ibid.*, 237, 18—28. cf., also, *PA* 238, 8 f.
27. *Ibid.*, 234, 6—10.
28. *Ibid.*, 239, 8—16.
29. *Ibid.*, 267, 29—31; 240, 12—14.
30. *Ibid.*, 172, 17—18.
31. *Ibid.*, 237, 30—38.
32. *Ibid.*, 258, 29—259, 13.
33. *Ibid.*, 177, 12—28. Husserl, by the time of the writing of Chapter XIII of *PA* undoubtedly regarded the abstract algebra of universal arithmetic as definitive of numerical arithmetic and not derivative from it. If so, "clarifying" the numerical signs of the latter by indicating their conceptual bases would not clarify arithmetic. The latter could be done only by a similar founding and rendering intelligible of the algorithmic procedures of calculation. The impulse in both instances is similar, if not identical. The only question is if Husserl, at the point of writing Chapter XIII, regarded the "art of arithmetical knowledge" to pertain to calculation or to the variables involved in it. In either case it would seem to be a clarification of signs. See Willard's discussion of Husserl's manuscript of 1890, "On the Logic of Signs (Semiotic)," (*LOK*, pp. 112—14), and of Husserl's position at the time of his writing Chapter XIII of *PA*. Willard concedes that nowhere, in that chapter or in any other of *PA*, does Husserl actually state that the key to such "arithmetical knowledge" is not to be found in concepts, whether "genuine" or "symbolic." The basic point is sustained by references other than to Chapter XIII (*LOK*, p. 115).

CHAPTER IV

The Symbolic Totalization Of Sensible Multitudes

In the chapter preceding his discussion of the symbolic representation of number in *PA*, Husserl explores "the symbolic representations" of sensible multiplicities.[1] These analyses parallel those following. However, the accounts *only* parallel one another. The pre-given or pre-totalized sensible multitudes are not abstractive bases for number concepts. Rather, Husserl demonstrates that the problems encountered in the realm of sense are analogous to those encountered in that of the formal. His prime objective with regard to the former is to account for how that which is "not directly given" but given "indirectly through signs," may also be understood in light of the mechanisms of "actual . . . representation."[2] Husserl cannot but seek to understand that which eludes the immediate grasp of consciousness in terms of that which does not.

This chapter is an excellent intersection of late nineteenth century thought at which to stand. It is replete with discussions of actual and inactual representations, the fusion of individual entities into a unitary Gestalt characteristic, and of the relation between these notions. Husserl acknowledges his indebtedness to Brentano and Meinong and Stumpf, as well as his lack of such to Christian Ehrenfels (i.e., to his article of 1890, "On Gestalt Qualities"). He declares that as far as he can see, "no one before" Brentano "had fully grasped . . . the eminent significance of the inactual representation for our entire psychical life." Yet he simultaneously declares his divergence from Brentano's definition of such representations. Husserl wishes to distinguish inactual from "universal" representation. It is here he argues that the "universal representation, 'a man,' " cannot serve as a "symbolic representation" of the determinate man, Peter. In order for it to serve as an inactual representation of him there must be the addition of "further characteristics" peculiar to Peter. Until this is accomplished, the universal representation is an inadequate "surrogate" for the actual representation of the individual.[3]

The Sensible Individual as Modified Multitude

Totalities of contents are convertible with the totalizing psychical act of "unifying and 'extractive' interest" which embraces them. Husserl contrasts these seemingly relationless wholes with those like the rose, the relations of which are physical or primary. By implication, these wholes would not fly apart in the absence of a unifying psychical act. The "well-characterized" relations which bind their *relata* are themselves contents, not acts. Yet, Husserl argues that such content relations are not always immediately discernible. Some, at least, may be seen only via an analytical process in which the psychical totalizing act is employed in reverse. Though Husserl does not make this explicit, he seems to imply that the psyche fully appropriates these wholes only through their deconstruction and reconstitution. Only in the course of this process do the content relations give themselves as "the more."

This whole might be characterized as one that "presents itself" to an "extracting" interest as the "unified intuition" of a whole already totalized. Its unity is not that of a psychical collective connection but of an autonomous content relation. This description might well be given of the rose or any similar sensible entity encountered. But it is that given by Husserl of the encounter with certain sensible multitudes.

Yet Husserl no more than so characterizes the sensible multitude than he contrasts the experience of it with the "individual sensible thing" (*Einzelding*). Whether or not it is his intent, his discussion of how sensible multitudes and individual things are to be distinguished also clarifies the similar way in which he characterizes both.

An analysis of the latter "after the fact," Husserl writes, discloses that it is a "multiplicity of parts" (*Teile*) and, namely, of "properties" (*Eigenschaften*). The constituents of the sensible multitude are not given as properties of the multitude but as "*partial intuitions separated* for themselves." In given circumstances, these partial intuitions may "attract" (*lenken*) a "dominant and unified interest to themselves."

The "representation of physical parts" can, Husserl acknowledges, also enter into the "representation of the thing." But in the case of the latter, as in that of the rose, attention rests on the relation of such parts with the whole, on their "belongingness" (*Angehörigkeit*) to it. It is this relation of belongingness that "makes" even the physical part a "*characteristic*" of the whole.

In contrast, the parts of the representation of the multitude *qua* partial intuitions, are not merely "separated for themselves" but "count" (*gelten*) only for themselves and not as "characteristics of the whole." Whereas analysis of the parts of the sensible individual only throws their inexorable relatedness to their whole into greater relief, when the parts of the sensible multitude are "intuitively separated" their "relation within the intuition of the whole recedes" (*zurücktreten*).

While the parts of the sensible multitude *qua* "unified intuition" are by no means united by the arbitrary totalizing act, they do nevertheless bear more of a resemblance to those of totalities than to the unities which are sensible individual things. Husserl nevertheless interprets the latter in terms of the totality as he does the sensible multitude. In this passage he refers to the "individual thing" as a "multiplicity" of parts and properties. The parts and properties analyzed ultimately manifest their physical relatedness to one another and to the whole only upon totalization.

In the case of the sensible individual thing where a "sharp separation" (*Trennung*) of "partial intuitions is lacking" (as it is expressly not in that of sensible multitudes), Husserl insists that it is wholly a function of the direction of interest as to whether it is considered a "multitude" (of, e.g., branches), or an "intuitable thing" (e.g., a tree).[4] This relatively cursory comparison of the multitude and the individual thing takes on far greater significance in light of Husserl's subsequent treatment of the individual sense object as a series of aspects in his "Psychological Studies for an Elementary Logic" of 1894.

THE SYMBOLIC TOTALIZATION OF THE SENSIBLE MULTITUDE

Interest having been solicited by the parts of this multitude (which, as the object of an "extractive" interest like the individual content, is a "unified intuition") the psyche's "primordial intention," Husserl writes, is the "formation (*Bildung*) of the representation of a totality." He leaves no doubt as to what the means to such a formation are. The representation of this totality (like all totalities) is formed by the grasping of each partial intuition for itself, and by the grasping (simultaneously, it is presumed) of it as unified together with the others.

The psyche no more encounters a group not of its own making than it summons its totalizing strength to fashion it in its own image. This is not, Husserl emphasizes, a mere inclination but its *primordial* (*ursprünglich*) intention. In the regard of the psyche, the multitude has little integrity of its own. It is but material to be deconstructed and reconstituted as a totality for which the psyche is responsible.

There are what Husserl terms "considerable" (*erheblich*) multitudes in relation to which the "intention (to totalize them in the manner described) ... lacks a corresponding mental productive-capacity" (*Leistungsfähigkeit*). While these multitudes are not so large as to preclude the "successive individual apprehension of the members" their "grasping-together collocation" is impossible due to their size. This holding-together of the individual contents is essential to the *actual* grasp of a group. Since in cases of considerable multitudes we "still speak of a 'multitude' (*Menge*) or a 'multiplicity' (*Vielheit*)," Husserl observes that these "can obviously only be seen in a symbolic sense."[5]

While it seems that this conclusion is accurate, given Husserl's framework, enigmas persist. Not the least of these springs from his insistence that "a sensible multitude ... presents itself to the 'extractive' interest, above all, as a unified intuition, as a whole." Husserl reiterates this point from the side of the subject when he writes that, in the encounter with such a multitude, "one glance (*Blick*) is sufficient" for us to judge that such is a multitude of, e.g., men or stars.

This *single* glance is reminiscent of the *one* act which embraces all the members of the totality. The actual grasp gives us, according to our "earlier analyses," writes Husserl, an "actual (*wirklich*) representation of a multitude" in that it "represents each member for itself and together with all the others." It was, he observes, *only with reference to* "*this form* of the psychical connection" of the "individually grasped contents" that the "names 'multitude,' 'plurality,' 'totality,' etc., acquired their meaning."[6] Husserl here clarifies the question as to whether the group which is a function of the totalizing act is to be called only a "totality." It is not, for whatever its name the group is intelligible only with reference to the *psychical* mode of connection.

This further illuminates the assertion in *BZ* that the psychical connection is the "psychological precondition" for all others. The actual totalizing maneuver of the psyche is clearly, in Husserl's view, the paradigm for all relations and, to say the same thing, for all collections. Whatever the collection, it must be appropriated by the psyche in this mode. It is clear in Husserl's description that a multitude has been seen, and that on this basis the judgment has been made: "A multitude." And it is equally clear that such a judgment, and its perceptual basis, are not without their respective meanings. Yet the one glance is not an *actual* grasp of the multitude. Still, it seems to present something similar.

This glance and the actual totalizing act cannot be identified by supposing that the latter could have occurred instantaneously in the former. If, Husserl argues, this actual grasp can be *consciously* accomplished with respect to "not more than a dozen elements," it beggars the imagination to suppose that it could be accomplished with, say, one hundred in a moment.[7] "It is indubitable," he concludes, that "the concrete representation of the plurality is in no way actual here." Therefore, "the subsumption under the general concept of 'plurality,' which is given with the application of the name 'multitude,' could only result from the symbolic way."[8] That with which the representational psyche is preoccupied is *either* a content actually in it, or symbolized by one that is. There is no third way.

Husserl's problem is this. The one glance grasp is that of a multitude of sensible objects. It makes sense to say that one sees "a multitude of men" in certain circumstances. However, the name and concept of "multitude" are only meaningful relative to the notion of their actual grasp. Here there are too many men for this grasp to be possible. This presentation of a multitude cannot then but be inactual or symbolic. What "is now the question," Husserl correctly states, is "where lies . . . the ground and 'foothold' (*Anhalt*) of the symbolizing?"[9] The multitude must be presented via "something" immediately given to consciousness which would stand for the actual grasp which is impossible due to the limited "productive capacity of our minds." This something must be graspable itself in one glance if such is the way in which the multitude is presented. It must be sensible, indeed it must somehow be "part" of the presentation of the sensible multitude itself.

The general movement of Husserl's analysis is similar to that in his treatment of arithmetic. There he began by assuming that psyches represent groups that can only be presented symbolically due to limited cognitive capacities. If these capacities were not limited, the symbolic presentation of large groups would not be necessary. Concepts of extremely large groups *are* had, however, and had only in virtue of symbolic capacities. Husserl's question at that point was, similarly, how are these concepts given, and what justifies the symbolic operations by which they come to be given? In the representation of large sensible multitudes, he acknowledges, as well, that the multitude is given and given symbolically. The philosophic question is: What is the symbol, and how can it present a great multitude in one glance?

Husserl concedes that the multitude is *not* apprehended in just one glance, its grasp is "not precisely momentary." More correctly, the eye "wanders" over this object and that, and consequently a "small group stands out" as actually grasped in virtue of the "apprehension and 'graspedness-together'" of these objects. He entertains the suggestion that this "simple rudiment" of the "entire process of collection" could be a "surrogate representation" for the latter.[10] This rudiment cannot suffice as the "foothold" for symbolizing, Husserl concludes, because it presumes what it is supposed to explain. The multitude has already been grasped as such, and this rudimentary totality is but a part of the larger group.[11]

Since this considerable multitude cannot be grasped in one act which, in attending to one content, holds all the others in view with it, Husserl seems to suggest that perhaps it can be successively apprehended in one act. This would provide some semblance of an actual grasp of the group and could also serve as a symbol for it. In demolishing this suggestion he also provides an illuminating discussion of time consciousness, the general structure of which persists in his later analyses.

In this passage Husserl asserts that the successive apprehensions of the members of the multitude cannot be held "together in once act," and he reads this succession from the side of the subject. It is all the individual *acts* of apprehension that the psyche is incapable of retaining. "Only a small number of such," he writes, "remain at a time in sharp distinction within the sphere of the colligating activity. During (this activity) new members are grasped and become added on, others, again, recede (*entfallen*) which were extracted earlier." This is only to say, Husserl clarifies, that the apprehending acts "fade way (*verschwimmen*) increasingly in the background of consciousness and finally disappear (*verschwinden*) altogether. Nevertheless," he continues, "we possess a determinate concept of the unity of the entire process," and this is so even if we do not actually complete the successive apprehension of each member. The concept is of this completion via "*some* succession" (which one is a matter of "indifference") which "brings all possible members of the intuitable whole to apprehension...."[12]

It is past apprehensions which are incapable of retention if the multitude is of sufficient size, if the succession of apprehendings of its members is long enough. Its members can neither be grasped together in one glance nor in one successive act. Husserl therefore refers to these apprehensions in the plural because they cannot be considered moments of only one act. The unity or disunity of the totalizing act is a function of the number of contents.

While the successive grasp of these multitudes takes "the light years of the astronomers," this recognition presumes at least the concept of their completion in a determinate time. This concept cannot serve as a symbol for the multitude grasped in one glance for the same reason, Husserl argues, that the rudiment could not. This concept of a "running-through" of all the members also presumes the grasp of the multitude as determinate. Further, it is itself inactual. It would be pointless to substitute a less satisfactory inactual grasp of the multitude for that which is already had. What must be identified is the *actual* content of the representing psyche which is the foothold for this inactual grasp. The suggestion of the actual rudiment of this multitude as this foothold was rejected because, while it could be grasped in one look, it failed to clarify the inactual grasp in one glance of the entire multitude. Husserl concludes by observing that the concept of the completed successional apprehension of each member also "miscarries" as a "try" at finding a symbolic foothold, and on "similar grounds."[13]

Given these miscarriages, Husserl opens the way for his resolution of this dilemma which he initially sets forth as a "hypothesis."[14] There "must lie," he asserts, "in the intuition of the sensible multitude, immediately graspable 'indications' (*Anzeichen*) in which the character of the multitude can be known, in that they indirectly ensure the 'performability' of the process described above" (i.e., that of the successive apprehension of members of the multitude). "With these 'indications,' then, the name and concept of the 'multitude' could themselves be immediately associated."[15] These putative "indicators" or "marks,"[16] these "immediately graspable" symbols *in* the intuition of the multitude, perform as surrogates for the actual grasp of the totality (be it in one look or in successive totalizing in one act). While the latter process is by no means actual in the case of considerable sensible multitudes, Husserl seems to maintain that it is the successive apprehension *of a "multitude"* only in virtue of "immediately graspable 'indications' " which provide its context as, precisely, that of a *"multitude."*

These symbolic elements of Husserl's hypothesis are clearly counterparts to those of the totalizing act. He expressly states that the intention when confronted by sensible multitudes is to *totalize* them. This intention frustrated, the psyche accomplishes it symbolically. And this process becomes a struggle to approximate the actual totalizing process.

In comparing the sensible multitude and individual thing, Husserl asserts that the members of the former do not serve as "characteristics" for the whole of which they are parts, as the parts or properties of the latter do for theirs. He observes similarly that the "characteristics" which will provide the foothold sought cannot "adhere" (*anhaften*) to the "individual members of the multitude" because they must facilitate a grasp of the multitude *as a unitary whole*.

The "foothold" to which Husserl alludes is that provided by the event of the members becoming "fused" (*verschmelzen*) in such a way as to exhibit "an *immediately noticeable* (and) *peculiar character*." Reminiscent of the terminology of "totalities of totalities," he terms this global character of the whole a "sensible quality of the second order" or "quasi-quality." It is second order simply because it results from the "conditioning elementary relations" of the individual sensible members of the multitude. The graspings of the latter are acts of the "first order." These "quasi-qualitative characters" are "conditioned" and thus indicative of the elementary relations and *relata* conditioning them. Nevertheless, they are the "πρότερον πρὸς ἡμᾶς" *vis à vis* these conditions, the means by which that which gives rise to them is known.[17]

Husserl is emphatic that these characters or "figural moments"(e.g., "a column of soldiers, a pile of apples, an avenue of trees, a line of chickens, a flock of birds, a gaggle of geese, etc.") are not contrived theoretical constructions but depend solely "on the *evidence of experience*," and find expression in the language "of ordinary life."[18] They are, he insists, experienced as incorrigibly *unitary*: "We grasp the quasi-qualitative character of the entire intuition as simple and not as a collection of contents and relations."[19] It is not given as the "mere sum of the relations" but as "a whole."[20] Husserl drives this home by pointing out that we see the "variation of figure" or gestalt prior to noticing the change in the relations of the members of the multitude which "necessarily condition" it.[21]

It is this pattern or configuration grasped as a unity which provides the peg or "foothold for the association" with it of the "name and concept of the 'multitude'" (the latter indicating the possibility of the grasp of a number of entities as a unity in one act.)[22] Husserl states this more explicitly when he writes that the quasi-quality enables an understanding of how "in one look a momentary, if also entirely inactual, subsumption under the concept of 'multitude' results."[23]

While he is clear that it is the unitary figural moment experienced directly, he is equally clear that, as ordinary language testifies, gestalts are always *of* something (soldiers, trees, fowl, apples, etc.). As indicative, they "indirectly guarantee the existence of a complex of relations, and with it (the existence) of its founding '*relata*' (*Beziehungspunkte*)."[24]

At this point the question arises concerning the relation between this "indirectness" of the *relata* and their relations giving rise to the figural moments, and the function of the latter as "footholds" for symbolizing. It is clear that these unitary patterns afford the grasp of the multitude in one glance and thus provide the possibility for an analogous (though inactual) experience to that of the actual grasping of a multitude. It is not, as J. P. Miller asserts, that "Husserl seems to be suggesting that the sensuous group itself is not actually present to us at all, that it is merely inferred on the basis of the figural moment."[25] It is not that the sensible members, e.g., each strutting goose of the gaggle, are not present for inspection (though not all are simultaneously since the multitude is considerable). It is not that this gaggle is somehow separate from its individual geese. On the contrary, Husserl acknowledges the awareness of the conditionedness of the gestalt by its founding relations of *relata*. If this barnyard spectacle is observed, it is known immediately that it is a gaggle *of geese*. Or, if its members are not recognized as geese, a configuration is nonetheless seen *of* fowl, or "ambulatory entities," or whatever. Insofar as this pattern serves to present the multitude symbolically, it is not separate from the constituents and relations of the multitude. Husserl makes this evident in his treatment of the way in which the latter condition the particular configurations.

That which the symbolic presentation represents, as in the case of the symbolic presentation of number concepts, is the *actual presentation* of the same group. Husserl indeed allows that simple "external intuition" ("external" indicates that no successive apprehension of individual members has occurred) of the "*figural characters*" can become the "*symbolic representative (Vertreter) of the 'actual' representation of the multitude.*"[26] The primordial intention to totalize actually the multitude finds, upon its frustration, an inactual way to present the multitude to consciousness. And the foothold the latter intention finds to do this symbolizes *the original intent and its "actualization."*

The movement is essentially that of the symbolic resolution of the psyche's inability to totalize entities exceeding twelve in number. The only difference is that large sensible multitudes are encountered as already exhibiting a unity. These simply present opportunities for totalization. Insofar as they resist actual totalization, Husserl summons the symbolic capacity of the psyche to grasp them with the tools at hand.

Since the related constituents founding the figural moment are given indirectly (and this only in the manner just explicated), Husserl states that these "indicators" also indirectly "ensure the 'performability' of the process" of the successive apprehension of the multitude. This is emphatically not an activity occurring instantaneously when the multitude is seen as a unity. The two are convertible in actual grasping because the apprehension of the group all at once is not sundered by the successive apprehension of its individual constituents. In the case of considerable sensible multitudes the two are convertible only if their symbols in some sense are. Husserl's account of the psyche's appropriation of these multitudes is but an adaptation of the moments of the actual totalizing act. And he is therefore unwilling to jettison its successional moment. In this regard he reiterates his earlier consideration of the partial successive apprehension of members as a candidate for the "foothold" sought: "Above all, where we find intuitively separated parts within a unitary appearance, their 'entirety' (*Gesamtheit*), the whole, finally exhausted in one successive process of extracted individual apprehensions, we attain . . . *a well-grounded symbolic representation of that collection corresponding to the unitary appearance.* . . ."

At this point, Husserl introduces "certain salient characteristics" (*Kennzeichen*) which result from the "fusion of the partial contents or their relations." These have already been "associated" with the "concept of a multitude." This association is itself now associated with the "concept of such a process" (i.e., that successive process just described) and this

70   Chapter IV

provides a "bridge" for the "*immediate recognition* of a multitude" which is the object, primarily, of a "unified sensible intuition...."[27]

Whatever may be thought of Husserl's account (presuming this reconstruction of it is accurate), of his elegant and interesting descriptions of Gestalt qualities, or of his rather contrived association of these with the symbolic functions of the successive apprehensional process, his ingenuity and single-minded imposition of the totalizing act cannot but be admired.[28] It is clear that this was his intent. Whether, or to what degree, it succeeded is another question.

Husserl's inquiry into sensible multitudes and his method of analysis are of utmost importance for an understanding of his approach to sensibilia. His presentation in *PA* is a thorough-going attempt to appropriate the multitude which is simply given to the totalizing psyche. It is of no small import that Husserl considers the sensible individual thing in this context. For it is precisely the totalizing act by which Husserl analyzes perception of the thing in his "Psychological Studies for an Elementary Logic" of 1894.

NOTES

1. The term may be rendered "sets," but it seems preferable to translate it as such when the context and reference is clearly that of purely mathematical or symbolic, not sensible, sets.
2. *PA* 193, 7—12.
3. *Ibid.*, p. 193, n. 1.
4. *Ibid.*, p. 195, 9—20, and n. 1.
5. *Ibid.*, p. 195, 20—30.
6. *Ibid.*, p. 196, 14—29.
7. *Ibid.*, p. 196, 35—197, 10.
8. *Ibid.*, p. 197, 11—16.
9. *Ibid.*, p. 197, 16—17.
10. *Ibid.*, p. 197, 18—29.
11. *Ibid.*, p. 197, 13—198, 8.
12. *Ibid.*, p. 198, 25—34; 199, 14—17.
13. *Ibid.*, p. 200, 11—201, 4.
14. *Ibid.*, p. 201, 5, ff.
15. *Ibid.*, p. 201, 6—12.
16. *Ibid.*, p. 201, 13.
17. Both the quasi-qualities and their conditioning relations/*relata* are within the general "descriptive" sphere of consciousness. Yet it may not be too great an exercise of analogical imagination to see here a modified analogue of the "descriptive" psychological/"empirical" or "explanatory" psychological split providing the larger contextual assumption of Husserl's early work.
18. *PA* 203, 19.
19. *Ibid.*, p. 204, 19—22.

20. *Ibid.*, p. 206, 9—11, see also n. 1 on Stumpf.
21. *Ibid.*, p. 205, 14—17 ff.
22. *Ibid.*, p. 201, 31—32; 10—12.
23. *Ibid.*, p. 202, 8—11.
24. *Ibid.*, p. 201, 32—35.
25. *NPA*, p. 50.
26. *Ibid.*, p. 214, 2—4.
27. *Ibid.*, p. 202, 39—203, 18.
28. Hussel claimed priority to Ehrenfels in the discovery of "quasi-qualities." Spiegelberg seems to regard Stumpf as having had the most influence on later Gestalt Psychology. See his *The Phenomenological Movement*, 2nd. ed. (The Hague: Martinus Nijhoff, Phaenomenologica, vol. 5/6, 1971) p. 54. Although he does not discuss questions of influence, Gurwitsch's article, "Phenomenology of Thematics and of the Pure Ego: Studies of the Relation between Gestalt Theory and Phenomenology" in his *Studies in Phenomenology and Psychology* (Evanston: Northwestern University Press, 1966) is helpful. See esp. pp. 250—267.

CHAPTER V

The Intuitive Totalization of the Individual Sense Object

Husserl's "Psychological Studies for an Elementary Logic" (hereafter *PSL*) of 1894 is immensely rich in that, on the one hand, it continues and deepens many themes of earlier publications. Husserl apparently continued to work on many topics he intended to address in the projected second volume of *PA* until at least the year of this article. The latter undoubtedly reflects ruminations on issues suggested by, if not adequately addressed in, the first volume.[1] On the other hand, *PSL* contains analyses which endured and were elaborated in subsequent publications (the most immediate being the third and fifth of his *Logical Investigations*).[2]

Specifically, Husserl continues and elaborates one of his most fundamental convictions, i.e., that entities with which the psyche is preoccupied, though not themselves immediate contents of consciousness, are represented by contents which are. He develops an insight here that is only negatively intimated in *PA*. The sensible signs of arithmetic, and of mathematics and logic in general, normally function merely as "game pieces" (*Spielmarken*) within the symbolic play of mathematicians. But they may be *interpreted* as designating one or another conceptual domain, e.g., that of numbers, and their interpreter is the mathematician.

The relative opacity or referentiality of signs (whether, in the terminology of *PSL*, they are "intuited" or, *qua* "representing," are not) is not a function of them as contents, but of the "manner of being conscious" of them (this is the "central thesis" of the article according to Husserl three years later).[3] To view this solely as a discovery of *PSL*, as some commentators have, is to overlook what seems to be its clear adumbration in *PA*.[4]

Husserl's analysis of "representatives" (*Repräsentanten*) in *PSL* is a further meditation on his deep question of how "the intuitive contents of the calculating arithmetician's mind could possibly function, as they most certainly do, in the apprehension of non-intuited — and usually non-intuitable — numbers and number relations."[5] This question is motivated most primordially by his epistemic-ontic assumption of the psyche as one of "represented contents" (*Vorstellungen*), and his further assumption that these actually embraced contents are most fundamental and certain. Husserl's own formulations of the question reflect his mathematical-logical agenda.

Normally, Husserl observes, in the "stream of conceptual thought" the objects referred to, "the significative contents" (*Bedeutungsinhalte*), infrequently enter consciousness.[6] It is true in "everyday life" that "'spoken' representatives aim at (*zielen*) the intuitables (*Anschaulichkeiten*) of the external surroundings. . . ." Even when the latter are present, they are "stingy" (*armselig*) and "fragmentary" (*lückenhaft*) and often so "transitory" (*unfassbar flüchtig*) as to be "incomprehensible." These "intended (*intendieren*) intuitions" are therefore "inadequate" relative to the words by which they are intended.

The fact that our intentions find so little "fulfillment" (Husserl uses the term in the article), does not really "trouble us." We simply assume that the "intended (*meinen*) objects" in our discourse "provide the foundation" for the "series of words" referring to them.

This situation does trouble Husserl because it seems to threaten "the possibility of knowledge in general" and "scientific knowledge" in particular. The latter, unlike that of "everday life," is *predicated* entirely on symbolic or "highly inadequate thinking."

Husserl observes that mathematicians have carried on "disputes" for centuries concerning the "meaning of its (mathematics') elementary concepts and the basis of the validity (*Triftigkeit*) of its methods." Such disputes on these fundamental issues are in "conflict with the alleged thorough-going evidence of its procedures." This evidence of procedures is even less than "alleged" because, he asserts, it is "pure deception." *Evidenz*, by its very nature, resolves disputes. The presence of the latter is but the absence of the former.

This pure deception within mathematics undermines the possibility of scientific knowledge or, at the very least, renders it a "mystery." What is mysterious is how this mathematics and logic can underwrite scientific practice which arrives at "empirically correct results."

This is Husserl's problem. Its irresolution mocks all pretensions to rational mathematical community, since the latter is constituted by evidential procedures engendering a corresponding body of evidential mathematical knowledge — an "exact science." The community of scientific practitioners is also mocked since its procedures are essentially mathematical. Husserl is convinced that this problem, and the mystery its irresolution promotes, can only be resolved by a "more deeply penetrating insight into the *essence of the elementary processes of intuiting and representing*." This insight cannot but produce a "full and actually satisfactory understanding" of these matters.

The intuitive appropriation of the sensible individual in the course of this clarification witnesses to Husserl's unqualified devotion to the totalizing mode. There is no question that he regards the sensible object as a collection of characteristics, parts, facets, and the like. While the psyche is *not* responsible for this unitary whole, it *seeks* to become so.

Husserl adapts this totalizing mode to fit the occasion. In the initial instance of totalities in *BZ/PA*, the "actual" grasp — whether in the course of totalizing or at its conclusion — refers to the psychical capacity to hold each content in view *along with* its compatriots. Whether the grasp is successive or momentary, these two moments of it are convertible. In the case of the sensible multitude, the incapacity to perform this grasp more or less in one glance was convertible with the incapacity to perform it successively. Both moments of the actual grasp were presented symbolically. However Husserl leaves no doubt that the psyche's "primordial intent" was the actual grasp. It was the impossibility of the latter symbolized by the inactual grasp.

These two convertible moments of actual totalization are split on the rocks of the sensible individual. Its multi-facetedness is not graspable all at once in more or less one glance — *even* symbolically. But unlike the considerable sensible multitude, it *is* actually graspable, Husserl maintains, in one successive act. His resolution of this dilemma is ingenious and inaugurates insights pertaining to time-consciousness that remained axiomatic for him.

Further, it is in their collaborative appropriation of the sensible object that the totalizing and symbolic psychical modes become virtually identified. Up until this point, the latter took up where the former left off, always indicating it, nevertheless. In the case of the sensible object, the actual or "intuitive" grasp is *impossible* apart from the symbolic function.

The Sensible Group: Sufficient Context for Analyzing Intuition of Individuals

In investigating these elementary processes of the psyche to clarify symbolical activity in mathematics and logic, Husserl simultaneously effects an analysis of the sensible individual thing. This is accomplished through his employment of the totalizing act. Intuition is defined in terms of the latter, and the intuited object is an *immanent* object.

Husserl describes the structure of consciousness itself as a sort of totality. He writes that "the current total consciousness is a unity in which everything stands in relation with everything else."[7] The total consciousness at any moment is comprised of everything in it and is not, therefore, a totality generated from a yet wider totality of consciousness, and so on *ad infinitum*. Consciousness is ultimately predicated on sensible phenomena and composed of innumerable recompositions, distillations, and complications of these. It is a totality of elements received and generated, the products of an immensely complicated and nuanced collaboration of activity and passivity. As such, it is by no means static. New sensational experience is ever taking place, there is ever the emergence of new elements out of the old and out of the old and the newly received. Husserl thus speaks of the "current total content of consciousness," each member of which is a "part-content."[8] This unified, yet ever-shifting totality of the contents of consciousness is analogous to the totalities generated by the totalizing act. Fundamental to the latter as well is the possibility of adding yet one more member to the group extant at a given moment. These totalities clearly depend for their contents on the reservoir of the current "total content of consciousness." They are generated by the successive attending to now this, now that content of it.[9]

Here then is the intersection with the "lifting-out" or "extractive" interest of the totalizing act of *BZ/PA*. The latter separates contents of its choice from the background of all the other elements of consciousness that might have been noticed "for themselves." And it is here that Husserl locates the possibility of distinguishing "intuiting" from "noticing." The former is defined in terms of what seems to be the act of noticing, and noticing in this new arrangement becomes but a shadow of its former robust self in the totalizing act.

If, he writes, "to abstract . . . is to attend (*beachten*) [to something] for itself," then such is precisely what we do when we "extract (*auszuscheiden*) an absolute concretum from its more encompassing background, and . . . make it the subject of particular occupation."[10] If, for instance, Husserl continues, "I inspect this box, I attend to it especially, and only in virtue of the fact that I do does it come to particular consciousness for me and become an object of intuition." That to which one attends is intuited. Conversely, that to which one does not attend, though there may be a marginal awareness of it as background, is not intuited.

Husserl is at somewhat of a loss as to how accurately to characterize this box *qua* object of abstractive attention. It would not be quite right to call it an "*abstractum*," yet neither can it be defined as that which is not abstracted. To complicate matters further, once abstracted it does not appear abstract, but eminently concrete. Husserl suggests it is precisely this abstractive attentional act which "impresses the stamp of the concrete" on the object.

In another passage he more or less reiterates this definition of intuition in writing that "only what is noticed for itself can be designated as intuited."[11] The "unattended-to . . . background" is not simultaneously intuited. That which is intuited is that "just extracted" with which one is "particularly occupied," on which "I have an eye."

Having said this, Husserl interjects that this treatment of intuition in terms of one object is "*insufficient*." He points out that "often" the one object, in this instance a knife, "stands out" with several other partial contents "for themselves."

The analysis Husserl inaugurates here is reminiscent of earlier applications of the totalizing act to sense perception and of that of sensible multitudes in particular. The group in this example is grasped actually so there is no invocation of figural moments by which considerable multitudes are grasped symbolically. The group of objects which "stands out" does so as a unitary configuration of contents not totalized by consciousness. And it is this sensible *group* which Husserl implies *does* suffice for his analysis of intuition.

"This knife forms," he observes, "with the inkwell, pen-wiper, and pencil standing next to it, a group of objects pushed up against one another (*entgegendrängen*), each noticed for itself."[12] Husserl now defines intuition precisely in terms of actual totalization. He notes that "exclusive attention" (*Zuwendung*) may be paid to one member of the group, thereby "stamping it as the intuited one." Yet, even during such stamping, the "entire group continues to be noticed." Intuition is defined here as (exclusive) attention. "Noticing" (which earlier fulfilled the function now assumed by attention) is relegated to referring to that which remains in awareness more or less as background.

Husserl distinguishes what might be termed a more "proximate" and a more "remote" background, although he does not do so explicitly. It seems that while other desk accessories are noticed only peripherally to the exclusive attention to one of their number, there remains an even more remote background generated by the abstraction of the group in the first place. The immediate desk area must also be in some sense noticed, though not in that of the members of the group which are not intuited. However this may be, Husserl has sharpened perception of the differences in contents through a more precise definition of the terms designating them.

It is not only individual members of the group which may be intuited. The entire group, he contends, may be intuited "as a whole" through the "more or less equal distribution" of attention over its members. This distribution of attention is egalitarian because it varies in its "preferring" of each content. Because all members are intuited as a whole, even those not preferred by this varying attention at a given moment are more than simply noticed. This "less-than-exclusive attending" or "more-than-noticing" *is* the "intuition" of the members of the group as a whole. It is presumably conditioned by the size of the group and by the relative proximity of the members to one another. As Husserl points out in *PA*, the unity of any configuration, regardless the number of individual constituents, is threatened if there are too great of distances between the latter. Husserl's four desk accessories are "pushed up against" each other.

He thus dismisses the lifting-out of the *individual* object as *insufficient* as a model for the definition of intuition. It is the actual totalizing grasp of the group which is advanced as sufficient for this definition. Yet a problem immediately suggests itself. This may all be illuminative of how sensible groups stand out as unitary wholes. It may well sharpen understanding of the intuited member's relation to others which are noticed and of how all members may be intuited as a whole. Yet may not Husserl's definition of intuition within the context of the sensible group be simply accidental and not essential? And even if his apparent preference for this context for the analysis and definition of intuition indicates its role is more than accidental, what is the precise nature of this "more than?"

If the totalizing act is essential to Husserl's analysis of intuition, then it must be relevant and applicable to the intuition of *sensible particulars in and of themselves* and not merely as members of groups. But such an application seems most unlikely if not incredible. Yet it is precisely this which Husserl does in *PSL*. In so doing he inaugurates an analysis of sensible particulars axiomatic in his later works.

The Problem: Non-Convertibility of Simultaneous and Successive Totalizing

Husserl initiates his analysis of the thing in *PSL* by drawing a fundamental distinction between "contents" (*Inhalte*) and "things" (*Dinge*). Things, he writes, are "indeed not the actual (*wirklich*) contents of our representations." They are, on the contrary, "objective unities" (*Einheiten*) which are only "presumed (*vermeintlich*), merely intended (*intendieren*) contents."[13]

The structure of this brief remark is reminiscent of the formula in *BZ*, i.e., if an object does not lend itself to an actual grasp, it must be grasped inactually or symbolically. Whatever the actual contents of this representation, they must be *representative* of the thing. Via these proxies, the latter must be a "mere intention."

In many of the instances Husserl cites, the fact that a given entity is not grasped actually is an index of the subject's incapacity to do so. Its intention to grasp the entity actually is frustrated and is consequently realized only symbolically. But all of this is not clearly spelled out in this brief passage.

Husserl does elaborate later on this "thing." Once again, he contrasts that which is "actually (*wirklich*) perceived" or seen, the "actually present content of the perceptual representation," and that which we only "believe we perceive," or "think that we see."[14] He cites judgments of comparison as illustrative of this contrast before returning to the consideration of the thing. We judge, e.g., that the sides of the cube sitting before us are equal. In point of fact, he corrects, there is "lacking a corresponding basis" in actual perception for this judgment. What we actually see from any given perspective are unequal sides.

Since Husserl draws this clear distinction between that actually perceived and that which we erroneously believe we perceive, it is obvious that at least he is capable of discerning epistemological error. He describes those who are not, who partake of "perception in the popular sense," as having "natural consciousness" and engaging in the "non-reflective view." Humankind divides into the non-reflective, those who presume (*vermeinen*) "to possess as immanent object" what they "merely intend," and those who are "psychologists."

There is, Husserl reiterates, a possessed and incomplete "intuitive content of the perceptual representation," and a merely "intended, complete content." Another way of expressing the error of the non-reflective view, then, is to say that it "presumes" itself to be in possession of the "complete content" of whatever when it has but a part of it.

Husserl deepens his analysis of the thing by way of describing the false consciousness of it. This consciousness "believes," Husserl begins, "that it grasps the objective thing itself, this 'unified diversity' (*einheitlich Mannigfaltigkeit*) — as that which it (both) is and is intended (as) — in one glance, in one simple act of intuition." This belief "we know," he writes, to be "mere illusion." It is illusory because "only a small part" of that which "we here mean (*vermeinen*) to intuit, is actually (*wirklich*) intuited — only a few features of the factual content are so present" in the "one glance" as they "are intended, and as they actually coexist in the 'thing itself.'"

This description of the thing is consonant with Husserl's earlier characterizations of it. In both his analysis of the rose in *BZ*, and the comparison with the sensible multitude in *PA*, he treats the thing as a "multiplicity of parts." However, the parts are linked and mutually implicative in their "belongingness" to the whole in a way the members of the multitude *qua* "partial-intuitions separated for themselves" are not. There is a unity of a contentual, physical, or primary sort binding the parts of the thing more intimately than those of the sensible multitude.

The problem posed by the sensible individual is reminiscent of that of the sensible multitude in *PA*. The sensible thing is perceived as unitary. Yet all of its parts or facets are not seen at once. The constituents of

The Intuitive Totalization of the Individual Sense Object 79

the sensible multitude were seen all at once, but inactually. This fact motivated Husserl's quest for the unitary "representative" of all of the parts. There is no grasp of all facets of the thing at more or less the same time, actually or symbolically. Such would be the grasp of the "thing itself."

Once Husserl determined that there was no actual grasp of the members of the multitude all at once, he proferred the possibility of an actual successive grasp of them in one act. It is this strategy he employs with respect to the thing in *PSL*. "The complete content of the thing representation," he writes, "only becomes intuitable in a continuous course of contents through which are certain psychic acts." The latter, "which accompany the series of obtruding partial intuitions, identify those connected through mutual reference and — extending *through one continuous act* — elaborate the objective unity."[15] *This* "complete content" is not that "merely intended" of the "thing itself." The thing, Husserl states clearly, *may* be actually or "intuitably" grasped in one *successive* act. This completed grasp may be intended only on the basis of a partial completion of it. But this intention is of something neither incompletable nor impossible.

It would be impossible if the act was not unitary and continuous. This unity quickly splintered on the reef of the considerable multitude in *PA*. The intention to grasp the latter successively in one continuous act remained mere intention. The sheer number of contents eroded its unity as they faded and disappeared from consciousness.

This failure to hold all elements of a considerable succession together in one act was convertible with the incapacity to grasp them simultaneously in one glance. In both cases "mental productive capacity" was incommensurate with the number of elements of the particular group. In his treatment of the sensible individual Husserl explicitly states that the individual apprehensions of partial intuitions do *not* fade and disappear, but persist in their respective accompanying of each partial intuition "*through one continuous act*." His insistence on the unity of this act implies that the course of contents of the thing is of insufficient length to erode it. It seems to be in virtue of this fact that Husserl can assert that the "full content" of the thing can become intuitable. Unlike the considerable multitude, the thing must not be presented inactually. Perception must be the actual grasp of the perceived in and of itself, even if it is given successively.

But this actual or intuitable grasp of the thing *qua* succession presents a problem. The failure to grasp all members of a succession in one act is convertible with the incapacity to grasp them all at once, more or less simultaneously. Conversely, the capacity for the former ought to convert with and to that for the latter. Such is the case with the original totalizing act. The ability to see all members totalized along with that currently added expresses itself as well in its grasp of the completed totality. This is not so in the appropriation of the sensible individual, the successive grasp is not convertible with the simultaneous grasp. If the former is actual, what is to be made of this? Is the thing grasped *both* actually (*qua* succession of facets) and inactually (*qua* grasp of all facets at once)? Is it both intuitable and not? Or does the actual grasp of the succession function as a symbol for the "thing itself?"

Husserl does not resolve this dilemma at the conclusion of the paragraph under consideration. He simply reiterates that "ordinary perception ... is not intuition of the thing." This perception is presumably that termed "popular," "natural," and "non-reflective." Husserl thus appears only to be reiterating that it is sadly deluded and mistaken insofar as it believes itself to possess the thing itself. He does not even allude here to the intuition of its complete content *via* a successive apprehension. He simply asserts that if a "psychological interest is turned upon the momentarily present content, on the partial aspect just as it is, then we have an intuition of it."

Husserl's resolution of this dilemma is forthcoming. He resolves it through applying the structure of the totalizing act. But this is not a *mere* application because this structure must be *adapted*. And this requires a more penetrating insight into its successive and momentary facets, one Husserl has already adumbrated. This simultaneously affords further insight into the very structure and capacity of temporalized psychical consciousness.

The Resolution:
Successive and Simultaneous Totalizing as Continuous

Husserl has already laid the groundwork for the resolution of this paradox for intuition posed by the sensible individual. In his quest for a "foothold" for symbolizing the large sensible multitude in *PA*, one candidate is the actually grasped "simple rudiment" of the "entire process of collection." This rudiment stood out of the larger multitude in virtue of the eye wandering over this object and that. There is no doubt that this was a *successive* process. It follows immediately upon Husserl's observation that the apprehending of the multitude in one glance is "not precisely momentary."

Husserl invokes this description of the not-precisely-momentary grasp of the small group again in *PSL*. The intuition of the entire group of desk accessories as a whole has an element of succession in it. Its intuition is the "distributing" of attention over its members. This intuitive attention "varies" in its preferring of the individual constituents.

This succession is not of such degree in either instance as to preclude the grasping of all the constituents simultaneously. There is then an element of succession in even the momentary grasp of all the contents. Husserl resolves the dilemma of the actual grasp of the sensible thing on these grounds later in the article.[16]

He begins by reiterating the earlier critique of popular or natural consciousness, here ascribed to "natural men." Such persons, he repeats, possessing only "one aspect of the object, mistakenly believe that they "intuit the thing itself, this objective reality in which . . . all sides, all of the intuitable parts and characteristics coexist." He also reiterates his earlier critique of this belief by succinctly observing that "we intuit different sides of the thing from moment to moment, (but) in none (do we intuit), however, the thing itself." Even if the "complete content" is intuited, this is but the sum of the successively intuited contents. It is not the simultaneous intuition of all of them (as they "coexist," Husserl writes, in the "objective reality"). The "thing itself," as the correlate of this non-intuition, does not become an immanent object.

The tension between the intuition of the complete succession in one act, and the (non-) intuition of the constituents of this succession simultaneously, remains unresolved. What ought to be convertible is not.

Husserl's options are two. First, he can treat the intuited succession, *qua* "immanent" to consciousness, as a representative for the impossible simultaneous grasp of the thing itself. If the latter is the actual grasp of the thing, then *the thing* is *not* actually grasped, it is not intuited. Sense perception is then a mode of symbolizing. Insofar as sensible entities are the very foundations of Husserl's conceptual edifice, this conclusion seems to destroy them. The bases of actual concepts would suddenly themselves be *inactually* grasped. This is intolerable. The second and only alternative is to jettison the notion of the "thing itself." Or, rather, it is the *extension* of the "not-precisely-momentary" grasp to coincide with that of the completed succession in one act.

It is the latter option Husserl chooses. For he asserts, abruptly, that the "concept of an objective unity of parts and properties of such and such nature, which coexist independently of our consciousness . . . *has no place in the sphere of natural thought*." This concept is, he continues, but the "product of reflection on the thing representations of the common life." Husserl writes that "we commonly say that we have an *intuition* of visible *things*." If the intuition of the succession is all that is had, and if the "thing" *is* intuited, then the "thing itself *is*," he concludes, the "series of contents, internally interconnected, accompanied, and encompassed by certain psychical acts, which we experience, under normal perceptual conditions, in the case of 'the inspection of the thing from all sides.' " This succession is the "ultimate fulfillment" of the "thing-representatives." Husserl emphasizes this ultimacy by adding that there "remains nothing left over to be intended." If we seek the ultimate fulfillment of thing representatives such as " 'house,' 'tree,' and the like," we find it, he reiterates, "in a continuous course of intuited contents encompassed by one undivided act which endures continuously throughout the successive diversity of contents." In virtue of the intuitional encompassingness of the one enduring act, the succession of contents "is thus *immanent*" in it. Immanence is therefore an *accomplishment* of the act, not a given. That which is immanent *qua* "noticed" only is also a function of this act.

The totalizing act shines brilliantly through these descriptions. Husserl adapts the act to fit this situation by elucidating a latent dimension of the act. In the case of the sensible thing, the non-convertibility of the successive and (more or less) simultaneous grasp of contents is remedied by the stretching of the latter to coincide with the former.

This intuited "internally connected series of contents" is an adequate successor to the earlier thing characterized as a "unified diversity." This description of the succession as "internally connected" or "hanging together in itself" resembles that considered earlier in which Husserl speaks of the "partial intuitions" as connected by means of "mutual reference." These descriptions are also consonant with that of the sensible individual thing in *PA* (as well as with the discussion of independent and dependent parts of objects found in the first section of this article). When the "intuitively separated" members of a multitude collaborate and manifest a unitary figural moment or gestalt configuration, the manner of their collaboration is that of a "fusion" or "smelting" (*Verschmelzung*). They manifest a unitary appearance only through such proximate relation to one another. Husserl seems to regard the individual thing as such a fusion of parts as to lack what he terms "sharp separation" (*Trennung*). Not only are there no discernible gaps, but when these are effected through analysis they only accentuate the implication of the separated part in the whole. The "belongingness" of such parts means that they imply one another and the whole of all of them. They are each "characteristics" of the whole. This accounts for how a uniaspectival grasp of one such characteristic functions as that of one facet *of a whole*.[17]

It was Husserl's intent to totalize actually the sensible individual, to present an account of its grasp in and by one act. The demands of the represented object necessitated the conflation of the more or less simultaneous with the successive grasp of all of its facets. The object sensed must be susceptible of intuitive totalization; that which is immanent to the psyche must be accounted for by it.

The Mutual Implication of Intuiting and Representing in the Intuition of the Sensible Thing

All elements comprising the given totality of consciousness are immanent. Nevertheless, some are primarily so in virtue of being attended to and, hence, intuited. Sensible contents of consciousness are thus divided into those intuited and those only incidentally "noticed" or "sensed." All that is intuited is sensed, but not all that is sensed is intuited. All sensible contents of which the psyche is aware are either contents of it (sensed or intuited) or represented by contents that are. Those represented are obviously neither sensed nor intuited. The question in *PSL* is that of to which category their representatives (*Repräsentanten*) belong.

The answer is adumbrated by Husserl's treatment of the sign in *PA*. Arithmetical signs normally function as abstract and opaque game pieces. This functioning does not preclude their interpretation by mathematicians as designating the "objects" of one conceptual region or another.

In *PSL*, Husserl points out that the manner of being conscious when intuitively preoccupied with, e.g., a sound or a beautiful Arabesque, is different than that when, suddenly, the sound is understood as a word referring to something, or when "this beautiful design" is seen as Arabic meaning, "God is Great!" In the case of intuitive attention to the sound (being fascinated "with a peculiar timbre of the voice or a strangeness in pronunciation") or to the design (being affected "purely aesthetically") consciousness is preoccupied solely with the given object itself. However, when "the understanding lights up unexpectedly" and the sound and the design are understood as words, attention is no longer directed exclusively to the sound or design, though they continue to be heard and seen, but to that which they represent.[18]

That which is represented is not "in" consciousness as either sensed or intuited. Yet it is this to which attention is directed, and with which, Husserl writes, that "we believe ourselves to be occupied. . . ."[19] But this occupation of consciousness with that which is absent, "really," he remarks, "occasions astonishment."

Conversely, since we are *attentive* to "absentee" contents in virtue of the representative, we are not attentive to the latter. As a result, Husserl insists, it is *not intuited. Qua* transparent in its referential and vehicular function, it is only noticed or sensed. This conclusion follows perfectly given the pivotal role assigned attention in perceptual consciousness.

Those sensible objects in consciousness which, *qua* "pre-vehicular," are intuited in virtue of an attending to them, are only sensed *qua* significative. Their status is peculiar in that they do not simply recede into the background because attention is shifted from them to another object which is intuited. They recede insofar as attention is directed through them to that which they signify.

These respective manners of being conscious "usher" one another in and are thus "interwoven" with respect to the same entity. One, having ushered the other in, bows out and *vice versa*. They cannot coexist and remain two distinct, though intimately interrelated, modes of consciousness. Husserl insists that, "if anywhere, it is here that the testimony (*Zeugnis*) of inner experience (*Erfahrung*) is clear. . . ."[20] The earlier theme of the complementarity and distinction of actual and inactual grasp is continued here.

Employing language similar to that of *BZ*, Husserl characterizes the psychical experience of representing (*repräsentieren*) as one which "merely intends" its object. He writes that "merely to intend something is to aim, by means of some contents which are given in consciousness, at others not given, to mean (*meinen*) them, to point to them with understanding. . . ." So to do, Husserl states, is "to use, with understanding, the former (given contents) as representatives (*Repräsentanten*) of these (not given)."[21]

If that "merely intended" enters consciousness in either the strong or weak sense of immanence, then it is no longer represented. However, this object of the representative aim need not come to what Husserl terms "ultimate fulfillment," to the manner of "intuition pertinent to it." He defines such as a "fulfilled intention," and defines other gradations and nuances as well, including "provisional" fulfillment, "pure" and "impure" intuition, and so on.[22]

The "representative" mode is also employed by Husserl with respect to "objects" not merely contingently absent, but non-intuitable in principle. He cites geometrical idealities which, *qua* "goals of idealizing, i.e., conceptual processes, are *eo ipso*, non-intuitable."[23]

The "surrogate" figures and relations of geometry, whether actually constructed with compass and pen or only imagined, bear a "certain analogy" to these idealities and are intuited. However this analogy which "places" the intended figures and relations "before us" by no means renders them intuitable.

Similarly, "all conceptual representations which include evident incompatibles," such as the round square, are "necessarily non-intuitable."[42] Husserl does not regard the meaning of such expressions as a function of the possibility of their intuitive fulfillment. They simply "have an entirely determinate and well understood intention, but (they are) directed toward something impossible."[25]

Husserl reintegrates the representative function into the very successive intuition of the sensible individual. The upshot is that their distinction virtually dissolves, for intuition of it is impossible without this function. A given facet of the sensible object must, as a facet *of* a unitary thing, serve some representative function for the latter. Husserl first mentions the subject in drawing an analogy between sense perception and phantasy. He observes the way in which the phantasy representation of the "Borghese gladiator" serves as a "surrogate" for the "merely intended intuition" of the gladiator himself. Husserl recurs here to his earlier terminology of "actual" and "inactual" representations. The phantasy figure would be intuited if a "particular . . . psychological interest referred to the immanent content of the phantasy representation itself." Otherwise, the intention is directed through it *qua* representative.[26] Yet, "another inactuality" pertaining to the phantasy figure "enters into consideration." This refers to the fact that normally "a thing is represented through inadequate representatives — through a one-sided, more or less incomplete, 'aspect.' "

Since this passage occurs prior to the presentation of what is presumably Husserl's definitive view on the matter, it is unclear whether the "thing" referred to is that of the psychologists. It does occur after Husserl's allusion to the complete intuitive grasp of the thing *via* its successive unfolding.

At a later point following the extension of the definition of intuition to embrace the continuous, Husserl gives a somewhat longer exposition of the point in question — though not one either that is entirely unambiguous.[27] The location of this passage does count for something, although, as in *BZ*, Husserl does not always present his definitive position without subsequent reversion to its antipodes.

The "thing itself" is neither intuited nor intuitable. Now Husserl begins this passage by saying, "When we 'intuit' a thing. . . ." Perhaps his setting "intuit" off in quotation marks indicates that this is not true intuition. However, even in the intuitive grasp of the succession of contents, all aspects are not given with each aspect. Husserl has stated that intuition may be either of the momentary percept or of the entire series of such. Might it be that he refers here to the intuition of the thing (given one aspect of it), which is not yet a complete intuition of the whole thing — hence, only an " 'intuition' " of the latter? This also seems possible.

However a problem immediately rears its head. Husserl speaks of the intuited aspect which, presumably, serves as the "means of understanding with relation to the . . . thing" which is "merely represented." But this aspect is then the representative and, given Husserl's definition of the experience of such, *cannot* be intuited. To complicate matters, he insists that we are "also turned *intuitively* toward the actually seen (aspect.)"

This apparent paradox is resolved when the peculiar nature of the aspect *qua* representative is recognized. This facet is a connected aspect *of* the entire thing. Husserl asserts, therefore, that "here (in the representative aspect) *the two functions* (intuitive and representative) *are united*." The sensible aspect of the sensible object does "place before us" that which it represents insofar as it is a part or an aspect of the latter. In the case of the sensible object, these "two functions" do not "usher" one another in and out, but co-exist simultaneously.

Repräsentation and intuition coincide and interpenetrate in the appropriation of the sensible object. There is always something more and beyond that part seen, and the unseen is represented, indicated and adumbrated by the seen. Yet, since that seen is itself a facet of "the more," the latter is not absent in perception in the same sense as, for instance, the sensible object not immediately susceptible of intuition and represented by a word. The actually seen facet motivates further perceptions. Yet having the latter does not change the fact that the previously seen aspect is an aspect of the object in the same sense that they are.

Husserl was always keenly aware that there is no intuition of the sensible object if it cannot be grasped as such on the basis of one momentary facet. The latter cannot then be intuited simply as autonomous and anomalous. It must be seen as "presencing" the object *qua whole*.

Yet Husserl is *not* primarily concerned in *PSL* to emphasize the relative "absence" of the sensible individual but to stretch the definition and capacities of intuition as to render it not merely sensibly, but intuitively, present. In so being it is, he observes, truly an "immanent" object.

The actual and symbolic modes of the totalizing psyche remain preeminent in its appropriation of the sensible individual. But more than this, *Repräsentation* becomes *integral* to intuition. In the totalizing of the sensible object, these two modes of the psyche are virtually identified. Husserl recognizes early on that the intuitional grasp of consciousness must span more than the elusive moment. Insofar as retentions are *memories*, it may be questioned whether Husserl's insistence on the extension of the unitary act as far as memory allows is not artificial. Does not the "certainty" associated with immanence diminish in proportion to the fading of that which was grasped? Whatever the resolution of these issues may be, it is certain that Husserl, in finding succession at the very heart of the moment, did not regard temporality as inimical to the actual grasp of the object by consciousness.

This intimate collusion of intuition and the representative function continues in Husserl's account of the sensible object in his *Logical Investigations*. There, on the eve of Transcendental Phenomenology, *Repräsentation* becomes the means by which the intuited representation (*Vorstellung*) of the "transcendent" sensible individual is possible.

NOTES

1. See Miller's discussion in *NPA*, pp. 13—15. He cites Schuhmann's *Husserl-Chronik* (p. 43) as evidence that "it seems that" Husserl "still had the intention of completing and publishing the (second) volume as late as November, 1894" (p. 26, n. 66).
   References to Husserl's article, "Psychologische Studien zur elementaren Logik," are to the version published in the *Philosophische Monatshefte* 30 of 1894, pp. 159—191. Two English translations have appeared, the first by Dallas Willard in *The Personalist* 58 (1977), and the second by Hudson and McCormick in *HSW* (Willard, in his *LOK*, has acknowledged the latter translation to be the better of the two). These translations have been consulted, but most quotations in this chapter follow neither unreservedly.
   In *PSL* Husserl almost consistently terms "actual" (*eigentlich*) representations (*Vorstellungen*), "intuitions" (*Anschauungen*), and "inactual" representations, "representations" (*Repräsentationen*). The confusion instigated by these English renderings is obvious. In order to convey that Husserl's psyche is a "representing" (*vorstellen*) consciousness, this term will continue to be used. When at all possible, "representative(s)" will be used solely as a translation of "*Repräsentant(en)*." When "representation" is used as a rendering of "*Repräsentation*," the latter term will be included in parentheses.
2. See the introduction to the translation of *PSL* in *HSW*, p. 120; also Willard, *LOK*, p. 6.
3. Husserl makes this assertion in his "Bericht über deutsche Schriften zur Logik aus dem Jahre 1894," published in the *Archiv für systematische Philosophie* III of 1897, p. 226. See Willard's translation of it in the volume of *The Personalist* 58 (1977), pp. 317—18.
4. de Boer (*DHT*, pp. 16—17) is one who appears to regard it as an entirely novel discovery of *PSL*.
5. Willard, *LOK*, p. 6.
6. *PSL*, pp. 187ff.; Section 7, entitled, "Excursus Concerning the Psychological and Logical Meaning of both Functions and the Importance of Their Investigation."
7. *Ibid.*, p. 159.
8. *Ibid.*, p. 180.
9. Husserl presents an intrinsically dynamic consciousness, within which the particular psychic acts examined operate. This article emphasizes the inherent temporality of consciousness as well.
10. *PSL.*, p. 167.
11. *Ibid.*, pp. 180—1.
12. The word "*entgegendrängend*" might just as well be translated "compressed." While more literal translations seem less elegant, they nevertheless convey meanings which more refined renderings do not.
13. *PSL*, pp. 166—167. Husserl uses the terms "*wirklich*" and "*eigentlich*" more or less interchangeably, sometimes with reference to "representations" (when, often, the latter term obviously refers to their contents) and sometimes, as here, with explicit reference to the contents of such.
14. *Ibid.*, pp. 168—9.
15. *PSL*, p. 170.
16. *Ibid.*, pp. 177—9.
17. In the second formulation cited, Husserl replaces "unified," as the adjective modifying "diversity" (*Mannigfaltigkeit*), with "successive," and "unified" (*einheitlich*) now modifies the word "act." He thus emphasizes that the "unified diversity" of the psychologists is now to be understood solely in terms of the succession, and the totalizing act has

88   Chapter V

always been unified. Nevertheless, this formulation leaves the impression that of the two successions (of the contents, and of the acts accompanying and encompassing them), the latter is the stronger, in the sense of providing the essential unity. However this may be, Husserl assumes a fundamental "connectedness" of the contents in other passages which is not, *qua* "primary" or "physical" connectedness (to invoke earlier terminology), in any sense a function of the act.

Husserl also employs the same model in his account of melodies (*PSL*, p. 180). As one listens to a melody in an auditorium, many other noises intrude from without — unrelated human voices and the rumbling of wagons. We may "notice" these noises incidentally, Husserl writes, but do not (in that they are relegated to the background in virtue of attending to the melody) give them "any sort of particular attention — perhaps not even for a moment." As a consequence, as in his description of sensible objects, this "noticed" background has been "heard," but it has not been "intuited."

The analogy of the perception of the melody to that of the sense object is even more precise, in that the former also "plays" over a given duration of time. "The immanent contents" of the melody "differ from moment to moment," and therefore in none of them "do we grasp the melody itself." Yet, as in the case of the sensible object, "we do with good sense say of the melody we hear that we have an intuition of it." Husserl follows this assertion by observing that, "similarly, we commonly say that we have an intuition of visible things." He does not return to the melody following his discussion of the manner in which we "have an intuition of sensible things," but it seems likely that the account he gives of them is applicable, "similarly," to it.

18. *Ibid.*, pp. 182—184.
19. *Ibid.*, p. 187.
20. *Ibid.*, p. 184.
21. *Ibid.*, pp. 174—5.
22. *Ibid.*, pp. 174—77. Willard takes the intuitive "fulfillment" of the representative as an instance "where experience has transcended itself toward its object" (*LOK*, p. 16). It is his view (*LOK*, p. 17) that "intuition" signals the point where " 'the circle of ideas' is broken through . . . and transcendence of the flow of consciousness is achieved." The major ingredient in this accomplishment is "the ontological schema of potentiality and actuality. . . ."

Arguably, the real locus for the discussion of "transcendence" in this article is not "intuition" but "*Repräsentation*." For it is in the latter that the psyche reaches out to that which is not present "in" (immanent, i.e., "intuitive") consciousness via the sign which is. In this act consciousness is preoccupied, "astonishingly," according to Husserl, with that which is "absent" or "transcendent." In some cases, it is in principle so — non-intuitable. Intuitional "fulfillment" of the representative does not, then, make the object more "transcendent," or, as Willard has it, make "experience" more transcending of itself toward the object. Husserl's main concern with that which is not intuited is *how* to bring it into the immanent sphere. The psyche is interested in that which transcends it not because of its "transcendence" *per sé*, but because of its non-immanence.

Husserl is not terribly concerned in *PSL* with the issue of brute transcendence of sensible objects of consciousness. This concern seems to have emerged only later. Arguably, once Husserl thoroughly "immanentalized" the object in intuition, he became preoccupied with the datum of its "transcendence" of consciousness *qua* intuited. It seems that Willard neglects the fact (one which he acknowledges) that all of Husserl's discussion of intuition and *Repräsentation* in *PSL* occurs *within* the broadly *immanent* sphere.

23. *PSL*, p. 173.

24. *Ibid.*, p. 171.
25. While these are obviously insusceptible of intuition, it is Husserl's belief that, in *attending* to them, one is not attending to nothing. The intention is "directed toward something," albeit "impossible" (of intuitive fulfillment).

Husserl briefly makes mention in *PSL* of "concepts," "objects of concepts," and "contents of concepts." His discussion of intuition in this passage (*PSL*, p. 165), does not appear to be with reference to concepts as such, but only to that from which they are abstracted. In the first half of the article he does discuss relations of "evidential necessity" such as that of the intensity and quality of the tone. One does not change without an alteration in the other. These are what would have been earlier termed "content" or "primary" relations.

Perhaps, given the non-intuitive attending to that which is, in principle, non-intuitable (attention being the essential ingredient in intuition), and the perception of certain relations of necessity, a glimpse is caught here of Husserl's movement toward his doctrine of Ideas in the *Logical Investigations.* These ingredients are not sufficient to motivate the latter doctrine, yet they would not impede movement toward it either.
26. *PSL*, p. 171.
27. *Ibid.*, p. 184.

CHAPTER VI

The Totalizing Act as Mediator of the Ideal and Real

Husserl effected a decisive shift in his doctrine of concepts, leading him to reject entirely the immanent object, only two or three years following his masterful accounting for it in terms of totalizing intuition.

This shift was no less than a complete inversion of his previous understanding of the origins of concepts in psychical acts. It effected a total inversion in the nature of the concepts themselves in that Husserl now construed them as idealities appropriated by thought, rather than as concepts abstracted from it.

This inversion effected no less of the same in the function of the totalizing act. It was no longer what Husserl calls the "form of the collection" in the *Logical Investigations*. Rather, its totalizing of given entities became but the individuation of the ideal or specific form of the determinate collection, the number *qua* species. This act assumed the function of mediating the ideal to the real.

Husserl's construance of his "*abstracta*" as Ideas is not surprising given his descriptions of their anatomy in his *Philosophy of Arithmetic* of 1891. Indeed, his concepts were virtually Ideatic by that point, lacking it seems only some subtle change in intellectual lighting to be recognized as such.

A hypothesis then set forth in this chapter is that Husserl's conceptual doctrine of that period was very much like Jastrow's "duck-rabbit" popularized by Ludwig Wittgenstein.[1] Husserl's "continuous seeing" of these *abstracta* as concepts abstracted from psychical acts lasted for a number of years subsequent to *PA*. It is in this manner of seeing them that he is vulnerable to, and guilty of, charges of psychologism. It is maintained in this chapter that Husserl regarded himself as close to its brink from the perspective of his Ideatic inversion in the *Prolegomena* of 1900. On the horizon, however, of this psychologistic seeing was ever the possibility of the dawning perception of these concepts *not* as concepts of psychical acts, and therefore not as concepts at all but as Ideas.[2]

Husserl's "conceptual" project is virtually complete in his embracing the doctrine of conceptual idealism. For ideality is different, in principle, than the real, and Husserl consistently rejects any construance of Ideas as real particulars. However Ideas may be appropriated in thought by real persons, and this recognition constitutes Husserl's doctrine of "intersubjectivity" at this time. The thought-acts of an Archimedes and a Newton, e.g., are incorrigibly particular and transitory, neither over-lapping nor intersecting. Yet there is a certain community between them insofar as each appropriates the ideal circle in their respective geometrical reflections.

HYPOTHESIS: THE INTERNAL MOTIVATION FOR THE GREAT INVERSION

Husserl's attempt to arithmeticize mathematical analysis presumed that it and its auxiliary concepts requiring rigorous derivation were derivable from elementary arithmetic. The "exclusive foundation" of the latter was assumed to be the concept of number. According to Husserl, this concept is constituted by those of collective connection and something conjoined in a conceptual series determined by psychical acts. These "fundamental concepts" are simple, they are not emergents from a collusion of yet other concepts and psychical acts. They can consequently be clarified only by pointing to the concrete phenomena from which they are abstracted by psychical acts.

Husserl characterizes these acts as "resting" on the abstractive bases of concrete phenomena. While the latter need not be sensible, they are incorrigibly particular or factual. Since they constitute the foundation of Husserl's towering conceptual edifice, his fundamental category of this period is the particular.[3]

The totalizing act is the concrete phenomenon which is the abstractive base for the most fundamental and motive concept of collective connection. This act *is* the totality of entities it thinks together. Objectified by a reflecting act, the totalizing act provides the base upon which this abstractive act rests. This base is therefore not contentual.

Husserl subscribes to the classical recipe for abstracting concepts from particulars which prescribes attending to what is "similar" or "like" in the latter. However, he rejects this inductive assay in abstracting the concept of something from the representing act, and it is likely he rejected it as well in abstracting the concept of collective connection. In virtue of formalizing their contents, these acts are themselves already virtually formal. Husserl employs an abstractive recipe that is premised on particulars, but does not always require an examining of a number of them exhibiting like characteristics.

He recognizes that the concept collective connection is not sufficient unto itself for generating determinate numerical concepts. Indeed, this concept implies that one begin with "something" (rather than with it), and that once it ("&") is employed with something, it implies something subsequent to it as well. A *determinate* series of "&'s" and "somethings" may not begin, or end, with an "&." The concept of collective connection requires the concept of something to connect. This requirement indicates that the simple totality presumes representing acts. The concept of something is abstracted from the representing act and not from its contents because this act may direct itself to "anything whatsoever."

Since the concept of "&" implies a minimum of two *relata*, it is the progenitor of the concept of a multiplicity of concepts of something. The concept of the totalizing act performs the same function in the conceptual realm as its abstractive base does in that of particulars. While Husserl maintains that the concept multiplicity is convertible with that of collective connection, this does not imply that the former refers only to the latter *qua abstractum*. The concept of multiplicity refers also to something that might possess the concept collective connection.

Husserl is most interested in the concepts of numbers generated by determining the conceptual series of multiplicity and the subsequent naming of each determination of it. These numbers also have an abstract pole (the concept *qua abstractum* of e.g., "Five"), and a universal pole (something possessing the given concept).

The concepts collective connection and something derive directly from the particular psychical totalizing and representing acts. The concepts of multiplicity and of determinate multiplicities are the progeny of these two concepts of psychical acts and consequently trace their lineage to the latter concrete phenomena. Husserl explicitly traces the "all-embracing" character of these concepts to these wholly arbitrary and promiscuous psychical acts.

Since the totalizing act is eminently external to its contents and consequently only attached to them, the concepts deriving from it are likewise external to concrete collections and attached to them in a non-literal sense.

The major difference between these number concepts is that few are of totalizing acts which actually grasp their contents. These actual concepts are concepts of psychical capacity. Inactual concepts are those of totalizing acts which cannot be accomplished in fact due to the large number of contents.

These concepts derive from psychical acts in yet another way. Husserl regarded number concepts as in no way extant apart from their being *produced* by psychical acts "again and again." As productions, they also refer to the psychical acts which reflect, abstract, combine, and determine their constituent concepts. Husserl employs the psychical functions of attention, interest, and memory, so central to totalizing, in abstracting and articulating the spectrum of the concept. The latter comes into being only in virtue of these functions. Symbolic or inactual concepts are produced also by the psyche, but only by its employing prosthetic symbols. These concepts are "constructed" by the signs and operations devised and pressed into service by the finite totalizing psyche.

The totalizing psyche is pivotal to Husserl's early doctrine of concepts in the following ways. (1) The concepts collective connection and something are abstracted directly from psychical acts objectified by reflective acts. They are therefore concepts *of* acts. (2) Number concepts issuing from the combination and determination of these constituent concepts derive mediately from psychical acts. Their lineage is traced to the latter *via* their constituent concepts. (3) The entire process of generating number concepts requires psychical acts which reflect, abstract, combine, determine, and apply. Number concepts are therefore the products of these acts. (4) The laws of universal arithmetic or formal logic continue to "refer" to psychical acts, even after Husserl's rejection of their derivation from numerical arithmetic. They may be interpreted as regulative of numbers, but seem to be essentially so of thought.

Arguably, and paradoxically, some aspects of this doctrine not only threw the pivotal role of the psyche into question as Husserl pondered them, but also suggested an answer in the form of the ideality of concepts. These elements are (1) the *ambiguous* reference of numbers, and their constituent concepts of collective connection and something, to psychical acts; (2) the attendant hiatus between abstracted concepts, generally, and their abstractive bases; (3) the consequently problematic theory of abstraction; (4) the consequently problematic doctrine of the attachment of number concepts to concrete phenomena.

First, Husserl recognized virtually upon the writing of *BZ* that universal arithmetic was not derivable from the concept of number. On the contrary, the region of numbers was but one among many to which its various combinatory forms might be applied. This was due, at least in part, to his recognition that the variables implicated in these various combinatory forms are eminently susceptible of interpretation. These are formal and do not refer, in and of themselves, to anything in particular. Indeed, it made little difference in arithmetical practice whether number signs were even taken as signs of number concepts.

If a number sign is taken as signifying a number concept, then the latter is the universal concept, i.e., that of "any arbitrary set as such" falling under whatever concept of determinate multiplicity. The universal concept may be attached to anything, to any concrete group of the given determination of members.

Yet if this is true, then even number concepts do not refer in and of themselves to the psychical acts from which their constituent concepts are abstracted. They give no indication whatsoever of such lineal derivation. In this respect they are similar to the forms of universal arithmetic in that the latter do not refer in and of themselves to the region of numbers to which they may nevertheless be applied. Husserl initially recognized the affinity of these concepts with certain psychical acts. The latter also related themselves to "anything whatsoever" and therefore provided the possibility for the non-inductive abstraction of concepts. Nevertheless, this affinity is not sufficiently strong to bridge what is the complete lack of reference of these concepts to their putative progenitors.

This is no less true of their constituent concepts of collective connection and something. Once abstracted from their respective psychical acts, the latter are oddly amnesiac with regard to their putative origins in these acts. They are also attached to whatever in an external and non-literal sense even if they are understood as concepts of psychical acts. Yet if they *are* concepts of such psychical particulars, then they should be identifiable as such (as is, e.g., the concept of man once it has been abstracted from however many particular men).

Second. Even in the case of concepts such as man there is a certain hiatus or lacuna between the lowest universal level of the concept and the particulars which constitute its domain. This is also true of number concepts. Once concepts have been abstracted from particulars, once one has ascended to the inscrutable unity of the conceptual *abstractum* and begun the descent from it, one finds the transition to concrete particulars impossible to execute. The gap is not closed even if one adds a number of concepts together to designate a particular. If a concept had indeed been abstracted, then one should be able to complete the smooth descent through and from conceptuality to particularity.

Third. The non-referentiality of the concepts collective connection, something, and number in and of themselves to psychical acts, and the hiatus between concepts and their particular abstractive bases, calls into serious question Husserl's founding of arithmetic as well as his doctrine of abstraction.

He insists that abstraction does not effect an alteration in particulars. They are not changed in the least and remain in awareness even when attention is directed toward what is common to them. The hiatus between particulars and their concept is only filled by attention to what is common and inattention to the idiosyncratic in the course of abstracting. What remains unclear is the manner in which the universal derives from and is related to the particular. If the latter *is* rabidly particular, *how* may one even attend to the universal in it? The mechanics of the ascent from this "origin" of the concept is thrown into question, indeed, into obscurity. This obscurity is only complemented by that of the descent from the concept to its putative origin in the particular.

Fourth, Husserl's doctrine of the "attachment" of number concepts to concrete phenomena is a direct function of his conviction that they are abstracted from psychical acts. This doctrine carries within it the seeds of its own destruction. For if these concepts are applicable to *anything whatsoever*, then they give no indication of being derived from psychical acts. They may be attached to the latter, but this does not entail their abstraction from these acts *via* their constituent concepts anymore than their applicability to carrots entails that they are concepts of carrots.

If these concepts are not those of psychical acts, *then they need not be regarded as attached* to concrete phenomena. This doctrine of attaching was entailed only by the assumption that these concepts were those *of psychical acts*. If this assumption is suddenly dubious, then the source of these concepts is thrown into question, to say nothing of their relation to psychical acts and to concrete phenomena. If they remain external to concrete collections, then they are so in a different and as yet unspecified sense.

Husserl's entire project of founding concepts in concrete phenomena, and of building a conceptual edifice upon them, has become thoroughly enigmatic. Yet this enigmatic state of affairs simultaneously pointed the way which he would ultimately take. For if Husserl concluded that unitary concepts could be derived from neither particular psychical acts, nor from particularity in general, and that the abstractive doctrine of concepts must therefore be rejected, then it seems likely that he was motivated to effect the "great inversion" and embrace his own doctrine of ideas.[4]

There is no doubt that it is this doctrine which he espouses in the *Prolegomena* of 1900. Paradoxically, the way to it was pointed by views which *could* appear "psychologistic" only from the perspective of the new doctrine they motivated. This remains a hypothesis for the moment. It must be either confirmed or refuted by a tenable interpretation of Husserl's *Prolegomena* of 1900.

Confirmation: The *Prolegomena* of 1900

Husserl begins an especially complete and illuminating presentation of actual number "concepts" in a familiar way.[5] "If we 'bring home to ourselves' fully and completely what the number five actually is," if, he writes, we produce an "adequate representation of five, we will ... first ... form an articulated act of collective representation of any five objects." In *BZ* and *PA* Husserl would have pointed out that the relations between these objects were not content relations but convertible with the totalizing act. Here there is no mention of regression to the act.

Instead, he asserts that there is "intuitably given in it (i.e., this articulated act of collective representation of any five objects) — as its 'form of arrangement' (*Gliederungsform*) — an individual case of the above-mentioned (i.e., that of Five) species of number." The individual form of arrangement of the objects functions in more or less the same way that the totalizing act did, except that it is an individuation of a determinate numerical form *qua* species. Yet Husserl describes the agent as forming an act of collective representation of any five objects. This particular act seems to be responsible for the particular collection. Still, *it* does not appear to be the form of the collection as it was earlier. Its function has in some sense changed. Further confirmation of this hypothesis is found in Husserl's assertion that this form of arrangement is intuitively given, and so apparently given among the five objects themselves.[6]

In what immediately follows Husserl employs familiar terminology which nevertheless takes on different meanings within this different context of assumptions. He writes that "with regard to this intuited individual, we ... perform an 'abstraction.' " In this abstractive act "we not only 'lift-out' the individual, the non-independent moment of the 'form of the collection' but grasp the Idea in it." This grasping is the occasion on which "the number Five enters as species" into consciousness.

Attention is not directed toward the contents, but upon what Husserl terms the "form of the collection." This term seems synonymous with "form of arrangement" used earlier, for this form of the collection is also described as individual. As such, it is not independent from those entities of which it is the form. Its dependence upon them distinguishes it further from its antecendent, the totalizing act or collective connection. Husserl here uses the term "abstraction" with regard to the "lifting-out" of this *individual* form from its *relata*. Presumably, it is abstracted because it is non-independent of the entities for which it provides the form.

Yet this abstractive attention is not directed to this form only, but to the Idea exhibited in it. The latter is clearly identified with the species, the number Five, introduced earlier. Husserl does not speculate on the ontic status of these species or ideas.[7]

The abstractive act seems to perform two functions. It separates the dependent form from the contents, and then grasps the Idea of the form in this individual. Yet while this grasping is described as a moment of an abstractive act, the species is not described as abstracted. Husserl does specify that to which attention is directed primarily in this act. For when one grasps the "ideal species, which is (absolutely) *one* in the sense of arithmetic," attention is not directed toward the "individual case" of either the "collective representation as a whole," or "the form inherent (*innewohnend*) in it, although not separable for itself." Both the form and the total collective representation of which it is part are individual.[8] One directs attention to the Idea and ignores its particular case in the same way as one did toward the concept when abstracting it.

Husserl continues by asserting that this ideal species of the number "may even become objective (*gegenständlich*) in such acts" (presumably acts of grasping it). Consequently, he concludes, this species is "without any portion in the singular individuality of the real with its temporality and transcience. The acts of counting arise and pass away. It is not meaningful to speak of such in relation to the numbers." Numbers are objectivated as individual forms of collections in particular acts of counting. This fact entails, Husserl implies, that numbers in and of themselves are not "real individuals." The fact that they are objectivated *in* such acts also affords some insight into the new function of the latter. The particular acts of enumerating are the *means* by which these ideal species come to inhere in the particular collections for which these acts are responsible. As in *BZ/PA*, the totalizing act brings into being a concrete collection of whatever entities it chooses. In so doing, however, *it* is no longer the form of this collection. The latter *qua* individual is the *individuation* of the ideal species of the given determinate form or number of the collection.

The inversion Husserl effects is clear. Since the totalizing act is no longer the form of the collection, it is no longer the abstractive base for the concept of this form, for the concept of totality. And it is not clear that this concept is abstracted from even the individual form of the totality as a primary or physical relation. The latter seems to be but the individuation of an Idea which may be grasped in it. However, this complete change in function is not a complete demotion of the act. On the contrary, it is this act by which the ideal comes to dwell in the real. Yet the act counting apples and oranges *cannot* do so *except* insofar as it individuates ideal forms of arrangement. Husserl propounds the classical view that reality is unintelligible without ideality and that ideality remains sterile if not realized.

The psychical individuating act is *particular, it is real*. Husserl is firmly persuaded that the conceptual — which is now *ideal* — cannot therefore be derived from it. He states the distinction between his earlier and current position on this matter quite clearly in a later passage.[9] Psychology, which is the "natural science of psychical experiences," he writes, has as its investigative task the exploration of the "*natural conditions* of our experience." In its sphere of investigation are the "*natural (causal)* relationships of our mathematical and logical activities." However Husserl proscribes the employment of this psychology as a means of illuminating mathematics and logic proper, for these disciplines are not to be identified with the natural or causal realm. It is Husserl's unquestioning acceptance of the latter categories that impels him to distinguish that of the conceptual ever more radically from them.

The "*ideal relationships*" of our "mathematical and logical activities," including the laws pertaining to them, "form . . . a realm for itself." Husserl describes this realm as populated by "pure general propositions" which are "built up out of '*concepts*.'" And he emphasizes that these "are not approximate class concepts of psychical acts, but rather *Ideas* which have their concrete foundation in such acts."

Husserl rejects his earlier doctrine of concepts because their domain is not that of "empirical individuals," *among which psychical acts are counted*. "The number three" or the "truth named after Pythagoras . . . are . . . neither empirical individuals nor classes of such." On the contrary, they are "ideal objects which we 'ideationally grasp' in the acts of counting . . . etc." Husserl's point seems to be that the number three, e.g., is not a class concept because it makes no reference to particular classes of empirical individuals. The function of acts as "foundations" has altered radically.

This fundamental distinction between class concepts and Ideas, and the pure general propositions built up out of the latter, is clarified further by Husserl.[10] He writes that "arithmetical propositions (these include both numerical and algebraic propositions) are concerned with ideal individuals" which are the "lowest species," yet the latter are to be "sharply distinguished from empirical classes." These propositions therefore "tell us nothing about what is real, neither about the real things counted, nor about the real acts in which such and such indirect numerical characteristics are constituted for us."

Husserl elaborates on the similarity and difference of *arithmetica universalis* and *arithmetica numerosa*. "Arithmetical propositions" are not concerned in the least with "'what is contained in our mere number-presentations,'" he writes, but with "absolute numbers and number combinations in their abstract purity and ideality." The propositions of universal arithmetic, the "nomology of arithmetic," are "rooted in the ideal essence of the genus number." The last individuals which "come within the range of these laws are *ideal*, they are the numerically determinate numbers, i.e., the lowest specific differences of the genus Number."

While "arithmetically singular propositions" arise by applying "universal arithmetical laws" to specific numbers, Husserl insists that these have to do with numbers *only* insofar as they "express what is purely part of the ideal being" of them. He is emphatic that "*none* of these propositions" of universal or numerical arithmetic "reduces to one that has empirical generality...."

This ideal numerical realm of Genus and Species recapitulates the structure of its conceptual antecedent. It is still the *specific* numbers which are the ultimate or last individuals in it. These, and the propositions in which they are embedded, do not refer in the least to empirical entities, including psychical acts. This recognition, coupled with the attendant insight that such concepts could not then be abstracted from acts, arguably motivated Husserl to reject the abstractive doctrine and embrace that of Ideas.

He is keenly aware of the intimate correlation of abstractive acts and class concepts, on the one hand, and of idealizing apprehensions and Ideas on the other, as well as of their mutual exclusiveness.

Husserl states that one must "come to a clear understanding, (both as to) what the Ideal is in itself and in its relation to the real, how the Ideal refers to the real, how it can 'in-dwell' (the latter) and thus come to knowledge.[11] The fundamental question is," he asserts, whether the "true content" of the "ideal thought-objects ... reduces to individual, singular-experiences, resolves into mere representations and judgments concerning individual acts, or whether the idealist is correct when he says that *such an empiricistic theory* (*can*) *be stated* in nebulous generality, *but cannot be thought out*."

Indeed, this theory cannot be *thought*. He insists that "all attempts to reduce these ideal unities to real individuals" are absurd. They are "unthinkable" because this reduction entails the "splintering of the concept in any extension of individuals, without any concept that would give this extension unity in thought...."

One may well concur with Husserl's invective against this empiricistic folly but rightly protest that he himself never engaged in splintering concepts. Nevertheless, from the vantage point of the *Prolegomena*, Husserl seems to have regarded this absurd abyss as the logical conclusion of his earlier theory of the abstraction of concepts. The latter are arguably the "class concepts" which he now rigorously distinguishes from "ideal species." He rejected the abstractive doctrine because concepts, once abstracted, did not refer to their abstractive bases as they should have. Coupled with the hiatus between them and particulars, this led Husserl to see that concepts could not be abstracted from individuals. A theory which alleged this to be practicable must maintain that concepts *are* resolvable into empirical individuals. But this allegation is absurd because if what it alleges to be practicable actually was, then one could not even speak intelligibly of an extension of particulars.

Husserl also explicitly refers to the empiricist theory of abstraction.[12] He counterpoints the sciences of the ideal and of the real in terms of their "absolutely insurmountable difference." Included in this difference is that the extension of the universal concept is that of "lowest specific differences" for the first science, and of "individual, temporally determinate singulars" for the second. Husserl concludes that a "correct appraisal of these differences depends upon the *final giving up of the empiricistic theory of abstraction* which ... precludes the understanding of all logical issues."

One might again protest that the context of this assertion is mostly concerned with the contrast between Ideal and *inductively* derived concepts, propositions, and laws of the natural sciences. Therefore it has not been demonstrated that Husserl's criticism of the empiricistic theory of abstraction was criticism of his own earlier theory of arithmetic. Indeed, the pivotal concept of collective connection in the latter was not derived inductively (as that of something was not).

Husserl's earlier theory did not seem to suffer from the probability, uncertainty, and vagueness which Husserl associates with the "idealizing fictions" of inductive laws (and which are contrasted with the precision and apodicticity of logical and mathematical laws). Nevertheless, insofar as he became convinced that his earlier theory entailed a swift slide into empirical individuality, he arguably regarded it as entailing a simultaneous slide into inductive legality. Husserl rejects his earlier doctrine not because it had *actually* landed him in inductive relativism but because he perceived it as *tending* in that direction. The conclusions he so vehemently rejects are those that *might* have been his.

He does criticize a theory of abstraction that is not inductive but which involves the abstraction of concepts from particulars. It is in this passage that Husserl's most explicit and detailed reference to his earlier errors is found.[13] He advances the proposition that "not every law for facts arises from experience and induction." Not all knowledge "arises out of" experience "by the well-known logical process which goes from singular facts, or empirical generalities of the lower level, to lawful generalities." On the contrary, it is possible that the "fundamental concepts of logic are abstracted in psychological experience (*Erfahrung*) together with the purely conceptual relations given with them." Husserl does not give a detailed recipe for this abstraction, but his description of it is reminiscent of his early procedure.

The virtue of this procedure is that "what we find true in the individual case we recognize, at a glance, to be universally valid because grounded only in abstracted contents." Therefore, "experience . . . yields an immediate consciousness of the law-governed character of our mind." Since this does not require an assay of a number of particulars, "our conclusion is . . . free from inductive imperfection." Its character is not one of "probability but rather of apodictic certainty . . . not (one) of vague, but rather of exact delimitation, it in no manner includes . . . allegations having existential content."

Having advanced this position, Husserl rejects it.[14] He asserts that this ploy to avoid the plagues of inductivism "cannot suffice." The subject here is primarily that of logical *law*. However all that Husserl states concerning it applies equally to "fundamental concepts," which this theory alleges are also "abstracted in psychological experience." Indeed, the laws founded on these concepts are immune to inductive erosion only because the concepts themselves are allegedly derived non-inductively.

Husserl does not dispute that "our knowledge of logical laws, as a psychical act, presumes the experience of individuals, since it has its foundation in concrete intuition." What he strenuously disputes is the position that "concrete individual cases have the function of logical grounds, or premises — as if" the "existence of these singulars entailed the generality of the law." This point is reiterated. *Psychologically*, our intuitive grasp of the law (or of the concepts it presumes) may require two steps — a glance at the "individuals of intuition" and the "lawful insight" related to such a glance. Logically, however, there is only *one* step, and this is the "intuitive grasp" of the universal. Neither this grasp nor its content, Husserl writes, is "an inference from individuality."

Husserl thus describes "abstraction" as a form of inference. In abstracting the concept from the particular, the latter functions more or less like a logical premise. The concept is "derived" from and even motivated by it. However appropriate these metaphors and analogies may be, it is evident that Husserl categorically rejects the possibility of the derivation of conceptual generalities from particulars. And this rejection clearly extends to that derivation presuming to circumvent the pitfalls of classical induction.

The view once held by Husserl himself that "sums, products, differences, and quotients, and whatever else appears as well-regulated in arithmetical propositions, (are) nothing (other) than psychical products subject to psychical lawfulness," is rejected.[15] These are psychical products because they result from various operations on numbers, and the latter "arise from colligating and enumerating which are psychical activities." These activities are not jettisoned, but their function is altered. They are no longer producers of numbers and all that flows from them, but individuators of Ideas. Husserl observes that "in spite of the 'psychological origin' of arithmetical concepts, everyone knows it is a fallacious 'shifting-over' " (*metabasis*) to regard mathematical laws as psychological.[16] This is so simply because their "origin" in psychical acts now means something entirely different than it did in *BZ/PA*.

Neither fundamental logical concepts, nor the laws governing them, may be inferred from particulars. No element of particularity is found in their meaning. There is no basis then for construing them as *essentially* regulative of psychical activities. "The statement that 'a + b = b + a' states that the numerical value of the sum of two numbers is independent of their position in the connection, but it speaks nothing," Husserl emphasizes, "of anyone's enumerating and adding." This statement functions as a law, like any number of others in algebra or formal logic. But with respect to its content or meaning it does not say: "Every *psychical act* combining one number with another is equivalent to another act combining the same numbers, but in reverse order." The latter statement, Husserl writes, "comes in first *via* an evident and equivalent transformation" of the law. This is not an illegitimate transformation, but it is required before the law may be understood as regulative of psychical activities.

Husserl recognized early on that the *arithmetica universalis* with which he was most concerned was a piece of formal logic. He also refers in the *Prolegomena* to the "natural kinship between pure logical and arithmetical doctrines, which has often led to the assertion of their theoretical unity." Consequently, "an argument that is correct for logic, must be approved for arithmetic as well," and vice versa, for the two are "sister disciplines."[17] It is Husserl's practice to make an argument within the context of either formal logic or pure mathematics, and then indicate its relevance for the other.

Having described the grasping of the numerical species in the individual, he asserts that "what we have performed here, with respect to pure arithmetic, carries over entirely to *pure logic*." It is obvious that "logical concepts have a psychological origin," but he cautions that the "psychologistic consequence" seemingly entailed by this datum must be rejected. It must because these concepts "could have *no empirical extension*." They are not universal because their domain is not that of "factual individuals." Instead, they are "*truly* general concepts" the extension of which is "*exclusively one of ideal individuals, of genuine species*."

It is more difficult to get clear on the distinction between universal and "truly general" concepts in formal logic because its terms, e.g., "judgment," are *equivocal*. The latter term may stand for either class concepts of "mental states which belong in psychology," or for "general concepts for ideal individuals, which belong to a sphere of pure law." The dichotomous disciplines are again those of pure logic or mathematics and psychology.

While the term "judgment" is equivocal, Husserl leaves no doubt that its reference within pure logic is not to psychical phenomena. He asserts that in pure logic this term is equivalent to that of "proposition," which is not a grammatical but an ideal "unity of meaning."[18] The various types of judgments are not "titles for classes of judgments" but of "ideal forms of propositions." When the logician speaks of "every judgment," he does not mean every "act of judging," but "every objective proposition." Husserl's easy movement between arithmetic and logic is evident here, for he cites the judgment that "$2 + 2 = 4$," and asserts that its extension *qua* "logical concept" does not include various acts of making it at different times.

The same is true with regard to the "principle of contradiction" which is a "judgment concerning judgments." It is not specifically a law for acts of judging but for "*contents* of judgments," for "ideal meanings" or "propositions." Husserl emphasizes the distinction between the "*psychological* mode of treatment" (in which terms are class terms for psychical experiences) and the "*objective or ideal* mode of treatment" (in which terms stand for Aristotelian genera and species). As in BZ/PA, that to which words refer depends on that to which attention is directed.

The law of contradiction is constituted by two propositions, one the negation of the other, and a connection. The latter is also ideal. Husserl has thus come full circle concerning the origin of the collective connection. He writes that if the logician "specifies the diverse forms of connections of representations (which form) new representations," forms such as "conjunctive, disjunctive," and "determinative connection," then his discourse is not of "phenomenal, but rather of specific individuals."[19]

Formal logic is not intrinsically regulative of psychical activities. It is a theoretical and not a normative, or even a practical, discipline.[20] Husserl writes that "one need only to look in order to see that to the extension of this logical lawfulness *belongs only judgments in the ideal sense*." Therefore, "*not* one of the actual or conceivable *acts* of judgment" is included in it (although they "correspond, in endless diversity, to these ideal unities"). "We do not have to regard judgments as real acts" and, Husserl emphasizes, "in no case would they be the relevant objects" as far as formal logic is concerned. The view that logic refers to "psychical phenomena and formations" is a prejudice perpetuated by psychologistic logicians.[21] The construance of "presentations and judgments ... syllogisms and proofs ... truth and probability ... necessity and possibility ... ground and consequent" as psychical entities is motivated entirely by presupposition.

It is on the basis of his conviction that formal logic is ideal and regulative only of ideal entities that Husserl rejects both its derivation from psychical acts, and the attendant construance of it as a "technology" (*Kunstlehre*).

His polemical *Prolegomena* is a sustained and vehement attack on the position that the "essential theoretical foundations" of logic lie in the domain investigated by psychology.[22]

The latter view regards logic as related to psychology as "any branch of chemical technology" is related to chemistry, or as the "surveying of land is related to geometry." The relationship in each case is that of a "practical application" to its parent theoretical discipline. "However one may define the logical technology," Husserl observes, "we always find (that) psychical activities or products are indicated as the objects of practical regulation."[23]

The theoretical relationships between ideal entities in formal logic, and the laws pertaining to them, could only become normative in virtue of their "valuation" as normative. This valuation is simply the result of the intent to apply them as regulative of entities other than those ideal, and to regard them as normative for the various transactions between such entities. Only at this point does the practical interest arise, an interest in how conformity may be elicited from these entities to the given norms. The investigator of the ideal regions of formal logical entities may determine that all human thought ought to be subordinated to such laws, even as the propositions of formal logic are. He may even elect to formulate a curriculum by which students are instructed in logic, so as not to fall prey to invalidities in argumentation. Husserl has no complaint against this as long as one recognizes the transformations as being effected relative to the purely ideal nature of formal logic. The seduction is the forgetfulness of the descent from the ideal in the manner described. It is this forgetfulness which allows, according to Husserl, the "interpretation of logical laws as laws concerning psychical facts," interpretation leading to "essential misinterpretations of these laws."[24]

Husserl himself fell prey to this interpretation of formal logic.[25] The fact that he did is most intelligible when placed within the context of his early theory of abstraction. Husserl continued to speak of logic as a technology for some time after his recognition that *arithmetica universalis* did not derive from but applied to *arithmetica numerosa*.

This recognition did not immediately motivate the insight that numerical concepts were not derivative from psychical acts. Formal logic could regulate concepts from which it was not derived, even if they were derived from psychical phenomena, or any other particulars. Husserl regarded certain concepts of formal logic, e.g., those of various connectives, as concepts of psychical acts even after his recognition that the laws of formal logic were not derived from number concepts. Yet the latter continued to be derived from psychical acts — indeed, in part, from the same psychical acts — as did the concepts of connectives employed by logic. It seems Husserl continued to assume that fundamental logical concepts (e.g., "judgment") derived from psychical acts.

In any case, his rejection of the intrinsic applicability of formal logic to psychical phenomena arguably presumes his "great inversion." Concepts such as that of conjunction were fundamental not only to numerical arithmetic but to universal arithmetic and, therefore, to formal logic. If these concepts could not be clarified *via* abstraction from psychical activities, all of these moments of mathematics would be similarly affected. The stage was then set for the eventual rejection of psychical acts as the abstractive bases for these concepts. Whatever the order of insights, Husserl's psychologistic construance of logic as a technology was but the expression of a much more fundamental psychologism.

It goes without saying that the psychical activities and elements which figure prominently in the earlier generation of concepts would now be regarded only as data for a psychology of the apprehension of the ideal. And this is the position Husserl takes. It is now the task of psychology to identify "psychical circumstances which cannot be specified exactly," i.e., the "natural conditions" of experiences of "*Evidenz*" which are the graspings of idealities of formal logic. While the latter are by no means subject to the flux of time, the evidential experiences of such do "arise and pass away." Such psychical circumstances or conditions include a "concentration of interest" or attention, a certain "mental freshness," "practice," and a certain "preparedness."[26]

Whether psychology describes the evidential insight involved in the "emergence of a logical act of inference," or that in the grasp of the law of non-contradiction, the experience is a function of the idealities involved and not vice versa. It involves "no reference to psychical experiences or acts."[27] Again, Husserl has come full circle from his earlier theory in which these psychical functions were *productive* of concepts.

Indeed, from the Ideatic heights, the psyche appears not only particular but subject to "natural causal conditions." It is unclear whether Husserl, when writing the *Prolegomena*, meant that this posture of the psyche was only that imputed to it by explanatory psychology operating with hypothetical constructs.

Eugen Fink, in his article bearing Husserl's imprimatur, concedes that "an occasional over-emphasis of the ontological independence of the 'ideal' does occur" in the *Prolegomena*. He maintains that this excess was corrected in the *Logical Investigations* which propounded the " 'correlative' attitude" toward the ideal in response to that "naive thematic attitude" taken in the *Prolegomena*.[28] Fink does not characterize this "over-emphasis" as a mere didactic ploy to heighten the effect of its correction in the *Investigations*. It is not implausible to hypothesize, then, that Husserl's Ideatic inversion, entailing the change in function of the totalizing act from that of being the form of the collection to individuating this form *qua* specific, may have occasioned the "naturalization" of the psyche.

Yet even if Husserl did, for whatever length of time, regard the psyche as real from the lofty Ideatic perspective, he could not have forgotten that it was *the* particular among particulars. He could not have forgotten its tradition from *BZ* onward of accounting for sensible reality in terms of its totalizing act. Indeed, the elaboration of the representative mode in *PSL* was an account of the bestowal of meaning by the act upon an opaque sensible prospective "vehicle." The only difference by the time of the *Logical Investigations* is that the meaning is ideal in the full sense of the word.

It is through the collaboration of the ideal and the representing act, considered in the following chapter, that the real is accounted for as simply an *interpretation* on the part of consciousness. It is in this interpreting activity that Husserl's "conceptual" and "sensible" projects coincide for the consummate conquest of not only spatio-temporal phenomena but of the causal world itself. The latter becomes then not merely an "existent" that "does not appear," in Brentano's sense, one indicated by hypothetical physical concepts, but entirely an interpretation *in* immanent consciousness. One's own body is no less an interpretation of immanent sense data, so that whatever "reality" Husserl may have ascribed to the psyche from the Ideatic perspective is decisively dissolved. Consciousness is constitutive, now, of this reality.

It is therefore the totalizing act and the sensible project which triumph in the end. Although the interpretation of transcendent reality is accomplished with the help of Ideas, the representative mode of consciousness is more fundamental in that it was in place from *BZ* onward. It is this mode of psychical activity which appropriates meaning to account for the sensible object once and for all. This priority of the psychical act is witnessed to by Fink as well, in his promise that the over-emphasis of the "ontological independence" of the Ideas would be ultimately rectified or "gradually mellowed down" by a constitutive accounting for them.[29]

Notes

1. Ludwig Wittgenstein, *Philosophical Investigations*. (New York: MacMillan Publishing Co., Inc., 1968), p. 194e.
2. The doctrine of "propositions in themselves" was part of Husserl's intellectual ethos, and one would not be guilty of historical fabrication in suggesting that he simply virtually absorbed the doctrine by osmosis. Dagfinn Føllesdal has, since the publication in 1958 of his *Husserl und Frege*, consistently held the influence of Frege on Husserl to have been decisive, and quotes Husserl as having said so. This writer is in general agreement with Føllesdal that Husserl's adherence to the "*Vorstellung*-meaning-reference" distinction, and to the doctrine of a non-extensional logic of meanings, does not entail Mohanty's conclusion (in *his Husserl and Frege*) that Husserl had thereby effected his ideatic inversion.

The Totalizing Act as Mediator of the Ideal and Real 107

Føllesdal's assertion that "whether one accepts psychologism depends on how one conceives of meanings, as psychological processes or as 'ideal' entities," is correct to a point (see his latest published exchange with Mohanty in *Husserl: Intentionality and Cognitive Science* [Cambridge: The MIT Press, 1984], pp. 53—54). That point is the fact that Husserl held a *non*-ideatic doctrine of *abstracta* which he nevertheless did *not* understand "*as* psychological processes." He emphatically distinguished his concepts *from* the latter from *BZ* onward, although these processes were undeniably both the abstractive bases and the means of abstraction. Husserl was therefore psychologistic in his theory of *derivation* of concepts, but not psychologistic *in intent* insofar as he sought to separate concepts from the concrete in every possible way — and especially from spatio-temporal phenomena.

This chapter does not pursue this debate concerning influence. It maintains, rather, that there are sufficient *internal* grounds for Husserl's Ideatic inversion, in the same way that Jastrow's drawing provides sufficient grounds, if not motivation, for the perception of it as a duck or rabbit. This metaphor breaks down in that once Husserl understood concepts *as* Ideas, he could not see them again as concepts. This hypothesis of internal motivation charts a course between Mohanty and Føllesdal. Mohanty may well be correct that Husserl's concepts were essentially Ideatic by 1891, and Føllesdal may also be correct in holding that Husserl did not officially acknowledge them, or recognize them as such until sometime between 1894 and 1896.

3. This is not to imply that Husserl attempted to reduce number concepts, or the ranges of higher mathematics and logic, to particulars, sensible or otherwise. On the contrary, he is exceptionally scrupulous in demarcating facticity from the conceptual realm. In *BZ*, e.g., he anathematizes importing spatial and temporal references into algebraic propositions involving abstracted concepts. Indeed, this scrupulosity and rigor moved him progressively into the arms of his doctrine of Ideas in the *Logical Investigations*.

4. The phrase is J. N. Findlay's with reference to another great philosopher. See his *Plato: The Written and Unwritten Doctrines*. (New York, Humanities Press, 1974) or its popularization, *Plato and Platonism*. (New York, Times Books, 1978). In Investigation II of the *Logical Investigations*, Husserl dwells on these topics at length, and specifically attacks the role he himself gave to "attention." Our concern with the *Investigations* will be limited to their doctrine of the sensible object.

5. Edmund Husserl, *Logische Untersuchungen: Prolegomena zur reinen Logik*. (Halle: Max Niemeyer, 1900). Republished in *Logische Untersuchungen*. Erster Band. *Prolegomena zur reinen Logik. Husserliana XVIII*, edited by Elmar Holenstein, (The Hague: Nijhoff, 1975). Husserl published a second edition in two parts (1913 and 1921), both of which included changes reflecting his transcendental doctrine. All references here are to the first edition only (hereafter *LU*). The pages of the Findlay translation are given in parentheses. See pp. 169—70 (179).

6. Husserl describes this "form of arrangement" as being "in" the total *act* of the "collective representation of any five objects." The representing act has its categorical content, but as with its primary contents, this comes to be represented *only insofar as there is an act*.

7. Findlay translates "*Einzelfall*" as "instance." This is not incorrect, but it evokes what are perhaps unnecessary Platonic associations which Husserl may well have wished to avoid. Cairns also warns against the translation. It is rendered here as neutrally as possible so that whatever "Platonism" there may be in Husserl's doctrine can burst through on its own power.

8. Having cautioned against translating Husserl so as to evoke unwarranted Platonistic associations, one cannot help but note that his descriptions of the unity of the species, and of it as "in-dwelling" the case in which it is individuated, do smack of such.

9. Edmund Husserl, *LU*, 182f. (189f).

10. *Ibid.*, p. 171, 5 (180).
11. *Ibid.*, p. 188 (193).
12. *Ibid.*, pp. 177—78 (185).
13. *Ibid.*, pp. 74—75 (108—9).
14. In the forward to the *Prolegomena*, Husserl states that he has distanced himself from thinkers under whose sway he had been when he began to philosophize. In rejecting this doctrine he distances himself from Brentano from whom he acquired it. See, e.g., *DHT*, p. 77f.
15. Edmund Husserl, *LU*, p. 168 (78).
16. *Ibid.*, pp. 169—70 (179).
17. *Ibid.*, pp. 168—70 (178—79).
18. *Ibid.*, pp. 173—177 (182—84).
19. *Ibid.*, pp. 173—177 (182—84). cf., 244—246 (237—39) where Husserl observes that these "concepts" of the elementary connective forms have "correlative concepts" which are the "formal objective categories" of "Object, State of Affairs, Unity, Plurality, Number, Relations, Connection, etc." The latter are "married" to the former "categories of meaning" by "ideal laws." Predicated on categories of meaning are theories of syllogistics, on objective categories theories of pure pluralities and number. *Both* poles of these concepts are realized in the *act* of thought. The totalizing act in this scheme simultaneously realizes the idea of the determinate collection and that of itself.
20. See Chapter 2 of the *Prolegomena* for a complete presentation of these distinctions.
21. *Ibid.*, p. 167 (177).
22. *Ibid.*, pp. 50—51 (90—91).
23. See the fundamental distinctions Husserl draws in Chapter 2 of the *Prolegomena* between theoretical, normative, and practical disciplines. Given the fact he regarded logic as wholly theoretical, it is clear why he considered its construance as practical, i.e., as far from the theoretical as possible, anathema and absurd.
24. *Ibid.*, p. 78 (111).
25. During most of the period preceding *LU*, Husserl regarded mathematics and logic as a "technology" (*Kunstlehre*) guiding and regulating thought (see, e.g., the Forward to *BZ*). This was also Brentano's view, and Husserl was probably first exposed to it when he attended Brentano's lectures on logic in 1884—5.
26. *Ibid.*, p. 108 (134); p. 187 (192).
27. *Ibid.*, pp. 182f. (189f).
28. See his article, "Husserl's Philosophy and Contemporary Criticism" in *The Phenomenology of Husserl*, ed., R. O. Elveton (Chicago: Quadrangle Books, 1970), pp. 84ff.
29. The phrase is Mohanty's. See his article, "Husserl's Thesis of the Ideality of Meanings" in *Readings on Edmund Husserl's "Logical Investigations,"* ed. Mohanty (The Hague: Nijhoff, 1977), p. 78.

CHAPTER VII

The Ensoulment of Sensation: Triumph of the Totalizing Psyche

Husserl's great Ideatic inversion in *LU* intersects with what might be called the "great reversal" in the same work. The latter decisively draws the fangs of spatio-temporal reality, causality, and the probability associated with the inductive method appropriate to it.

The psyche's new function of realizing Ideatic forms of collections no doubt motivated Husserl to entertain the possibility of it performing a similar function for the sensible *relata* themselves. The form for the collection was no longer the act but the contentual realization of the objective species. It is likely that this inspired Husserl's rejection of sensible *relata* construed as merely immanent. His rejection of the empiricist's doctrine of concepts was attended by a similar rejection of their epistemology of sense objects.

Husserl attempted to account for the sensible content of consciousness in terms of the totalizing act from *BZ* onward. This act articulated physical relations of sensible individuals, and symbolically totalized sensible multitudes in lieu of its inability actually to do so. In *PSL* it accounted for the *actual* immanence of objects *qua* intuited, as opposed to that tacit immanence of those merely sensed.

Husserl's rejection of empiricistic doctrines in *LU* is anticipated by the fact that, from *BZ* to *LU*, he attacked the relativism he associated with spatio-temporal reality on two fronts. On the first, he progressively articulated and secured the conceptual realm against particularity, and on the second he simultaneously cultivated totalizing consciousness. The latter was accomplished by accounting for sensible contents as much as possible in terms of the totalizing act.

The need to account in *LU* for the sensible object as transcendent motivated Husserl's rejection of his doctrine of the immanent object *qua* intuited in *PSL*. He rejects it within the context of a thorough-going critique of the empiricist dogma of the sensible object as immanent. This "Lockean prejudice," as Husserl calls it, failed to give an account of the object in terms of conscious activity because it simply subjectivized the object as it was found in perception. Brentano, in Husserl's view, drew this prejudice to its logical conclusion. In importing the object of external perception into the immanent sphere as a configuration of sensations, he infected immanent consciousness with the doubt that should have remained outside. Husserl rejects the Brentanonian distinction between sensations and acts of sensing. He therefore regards Brentano as left with a configuration of sensations fraught with uncertainty and a passive consciousness.

Husserl was convinced from *BZ* onward that consciousness could be preoccupied with that not actually in it only by means of that which was immanent to it. This conviction inevitably disposed him to seek to account for the sense object *qua* transcendent in terms of representative activity. The refinement of the latter in *PSL* presented the perfect means of accounting for transcendence in terms of the realization of ideal meaning.

The representative mode required, however, immanent content that might function as the vehicle for the perception of the transcendent object. Husserl invokes putative sensations which, like the Arabesque of *PSL*, were opaque, indicative of nothing beyond themselves. Also required were meaning-conferring acts which could animate or ensoul these sensations with meaning. The result of the interpreting of these sensations by the act is the perceived object. As the product of meaning-imbued sensations, it is an *interpretation*. Here the interpenetration of representing and intuiting (now perceiving) in *PSL* is seen. The interpretation, the object perceived, *is* the sensational material *functioning representatively*. This interpretation is not a sign for some object beyond it that is "really" transcendent.

Husserl's two modes of attack upon spatio-temporal reality coincide in this doctrine of the sense object. Ideas imply nothing actual, it matters little whether objects exemplifying them exist or not. Their realization in representing acts interpretive of sensations *produces* the spatio-temporal world *qua interpretation*. Husserl thus decisively reverses the usual understanding of sensations as products of external impingments. Sensations are but immanent stuff which, when ensouled with meaning, yield an *interpretation* called the spatio-temporal world. The causality within the latter is but an interpretation constituted within the teleological processes of consciousness, as is the human body itself. Husserl secures Ideatic norms in *LU* as well as the consciousness apprehending them. The two means by which he has sought to subdue spatio-temporal relativity coincide.

Yet, his representing sensations seem insusceptible of inner perception. Husserl acknowledges that there is no awareness of them in the course of perceiving. While he maintains that they may be grasped adequately in inner reflection, it is unclear how this can be accomplished. If it is impossible to perceive sensations apart from interpretive acts, and *vice versa*, then all that may be given is the interpretation itself, perhaps shorn of its regard as transcendent. Yet if this is so, then Husserl seems to be left with little more than the immanent object of Brentano. Further, it must be concluded that these sensations and interpretive acts are but theoretical posits, demanded by representation but imperceptible. It is dubious whether Husserl's doctrine of sense objects as interpretations works on its own terms, regardless of whether the presuppositions motivating it are ultimately tenable.

The Immanent Object as Empiricistic Fetish

Husserl contends in *LU* that the empiricist epistemology of sense objects, to which he has himself more or less subscribed, is deluded. Empiricism claims to have clarified the relatedness of the object to consciousness. On the contrary, Husserl asserts, it has only obscured the way consciousness produces the object it perceives by merely importing it as it is given into the immanent sphere. The British empiricists, and Franz Brentano as well, succumbed to a fetishism of the object similar to that of those who naively assume it exists apart from consciousness entirely. The uncovering of the labors of consciousness which produce the object must then begin with the fetish itself.[1]

The most basic shortcoming of British empiricism, according to Husserl, was its failure to recognize that the word "idea" had more than one meaning. This failure led to its confusion of the "immediate object of perception" and the "representation" (*Vorstellung*) *of* the object or the "intentional experience." Since both are called "ideas," Locke confuses the represented with the representing, "the act with the intended object, the appearance with the appearing."[2] He consequently assumes, as do his successors, that "to have a representation" and "to experience (*erleben*) a content" are one and the same thing.[3]

Husserl himself always distinguished acts and their contents as a matter of principle. The latter nevertheless remained contents *of* representing consciousness. In *PSL* Husserl effected the activation of immanent contents in that the totalizing intuitive act *accounted* for them *as* immanent. The sphere of consciousness was thus divided into that which was merely immanent *qua* sensed or noticed, and that which was *actually* immanent in virtue of having been intuited by the act. Husserl has thus moved beyond the empiricists he criticized in at least attempting to *account for* the object in terms of totalizing consciousness. His was an eminently *immanent* object in *PSL*. In *LU* Husserl states explicitly that he rejects this "concept of intuition" propounded in *PSL*.[4] What he now regards as immanent are the "sensuous kernels" of the representing act and the "interpretive" (*auffassend*) act. The latter interprets or "means" (*deuten*) the sensations "objectively" (*gegenständlich*).[5]

As a result of the act of interpreting, there is an *interpretation*, which is the object perceived. The sensations and interpreting acts are not merely to be distinguished from the perceived but are "in principle antithetical" to it. Language obscures this antithesis, Husserl points out, because such words as "color" and "smoothness" and "shape" may refer either to "objective qualities" or immanent sensations. The latter are "representative" or "constitutive" (*darstellend*) moments of the perceived but are not themselves perceived. That which appears as objective or transcendent of consciousness is the result of an "animating" (*beseelend*) interpretation of these immanent moments. Since the same words are used to refer either to immanent interpretive acts and sensations or their interpretations, these antitheses may be identified or conflated.

This conflation hardens into what Husserl terms the "Lockean prejudice," i.e., that perceived objects are "psychical contents," that they are "real (*reell*) occurrences" *in* consciousness and are therefore *immanent*.[6] As a result, all that should be located in the transcendent object *qua* interpretation is placed instead in immanent contents. The upshot, laments Husserl, is that empiricists treat what are objective properties as "contents in the psychological sense," as immanent, as *sensations* only. This is also a fundamental confusion of perception and "psychological reflection." It is only in the latter, he maintains, that sensations are represented and may become objects for judgment. They are not what is perceived straightforwardly. When we represent or perceive a horse, Husserl insists, it is the *horse* (*qua* interpretation) that is perceived, not the sensations (which are experienced *qua* interpreted and thus represent it).[7]

Locke has thus made a dogma of an epistemological confusion. What is most regrettable about this in Husserl's view is that *it obscures the intrinsically active and necessary function of consciousness in the appearance of the transcendent object*. Those who remain adherents of this dogma are "bewildered," he writes, by the confusion of object with psychical content. In their bewilderment they overlook the fact that objects *of* which they are conscious are not *in* consciousness as if they were in a box. They are neither givens in the sense of natural realities existing independently of consciousness, nor in that of subjectivized versions of the latter. On the contrary, objects perceived are "*constituted*" by consciousness, by its active interpretation of sensations as objective. The definition of the representing of a content as objective in terms of its experienced sensations is, Husserl asserts, "a conceptual garbling which is still without equal in philosophy."

Husserl criticizes Berkeley and Hume for perpetuating the Lockean prejudice. Hume, for instance, also confused the having of a perception with the perceiving of an object.[8] In his critique of the latter's doctrine of the *distinctio rationis*, Husserl asserts that Hume "ran together" the perceived or apparent objective white sphere with its uniform coloring, and its subjective "sensational complex" of white which is not uniform. The uniformity of coloring of the perceived sphere is a function, Husserl holds, of the *interpretation* of its sensations.[9] The sensation white and the white that is perceived are antithetical, they "stand over against" one another in principle.[10]

This errant empiricistic dogma is no less extant, asserts Husserl, "in our day," and he reiterates the example of the sphere (which is red) as proof of the continued confusion of immanent sensations and objective properties of external things.[11] He regards Brentano as laboring under the same prejudice as the earlier empiricists.

"Phenomenon" is another term Husserl identifies as equivocal. On the one hand it is applied to apparent objects and their characteristics and on the other to immanent experiences of sensations and their interpretive acts. The consequence of this, again, is the mixing of what are two "essentially different" psychological kinds of division of phenomena.[12] These divisions are those, on one hand, of "experiences," which is divided into interpretive acts and sensations or "non-acts," and on the other hand, of "phenomenal objects," which divides into those that are "psychical" and those that are "physical."[13]

Husserl accuses Brentano of badly muddling these categories. Brentano correctly distinguishes psychical from physical phenomena, but defines both *as experiences* falling, respectively, into the categories of acts and non-acts. In doing this he makes the same mistake as the British empiricists. He treats as immanent the perceived object which Husserl insists must, *qua* interpretation, be understood as objective and transcendent. His "physical phenomena" are therefore merely immanent sensational configurations.

Brentano distinguishes sensations from acts of sensing, but Husserl regards this distinction as untenable. The experience of sensation is not to be sundered into act and object. In Husserl's view, Brentano's epistemology boils down to the confusion of objects interpreted as transcendent with immanent sensations. And if sensational *acts* cannot be separated out from the latter, then Brentano construes consciousness as eminently *inactive*. Immanent sensational configurations are simply there within it.

Husserl presses his criticism of Brentano further. He writes that when a house is perceived, it is perceived *as* an external object. This is possible only in virtue of the *experiencing* of "presenting contents" or sensations which, interpreted as representing a house, offer *the house* as external to consciousness to be perceived. Now this interpretation of sensations, like all interpretations, *may be incorrect* and the house, Husserl observes, *therefore may not exist*. If this is the case, we nevertheless remain in possession of "experienced sense contents," and he insists on our ability to reflect on these contents and abstract from our prior "meaning" of a house through them. It is unclear whether he believes that we may abstract from all vestiges of our interpretation of these sensations so that they *appear meaningless*, or whether it is only their interpretation as *transcendent* from which abstraction is made. In any case, it is Husserl's view that because this perception is of immanent contents, it is evidential.[14]

Brentano, in construing as immanent what he should have understood as transcendent (and therefore possibly non-existent), imported the "delusiveness" (*Trüglichkeit*) associated with the external object into the immanent sphere. Consequently he was forced to deny evidential status to immanent sensational configurations, and to fall back instead on the acts intending them as the loci of certainty.[15] Brentano therefore not only loses the certainty that should attend immanent sensations, in Husserl's assessment of his epistemology, but in so doing sacrifices virtually all conscious activity that might be associated with them (since Husserl rejects the distinction between *acts* of sensing and their sensational objects). Brentano is left with a theory of a consciousness inhabited only by inert sensations. Even the existence of the latter is uncertain. Husserl no doubt regarded this as the logical conclusion of the Lockean prejudice. It is Husserl's prime concern in *LU* to *account for* the object represented as transcendent or external to consciousness. And his tools for this task are virtually in hand. Ironically, it is his rejection of the empiricistic view of the psyche as the active abstracter of concepts and base for abstraction that occasions Husserl's far more radical activation of consciousness. The psyche's new function of realizing the Ideas or species of determinate collections in *LU* may well have motivated him to contemplate its engaging in similar action relative to the sensible *relata* themselves. The doctrine of Ideas transcendent of immanent consciousness (which had previously been abstracted from acts of the latter) may well have called for, in Husserl's view, non-immanent objects as well.

Yet Husserl's doctrine of the perceived object in *LU* is not as inimical to his doctrine of the immanent object in *PSL* as he might have assumed. The totalizing psyche prior to the Ideatic turn was anything but passive. In *PSL* it *accounted for* the *immanence* of the object via the interpenetrating of its totalizing and representative modes. There the intuition of the sensible individual is impossible apart from the representation of facets not intuited by that one which was. Representation is integral to the intuition of the sensible thing as unitary. The totalizing and symbolizing modes of the psyche always collaborated prior to *PSL*. Symbolizing was but the prosthesis of actual totalizing. But prior to *PSL* the presence of one mode rendered that of the other more of less superfluous. In the actual or intuitive grasping of the sensible individual in *PSL* they are mutually implicative and equally necessary simultaneously.

Representation founds intuition or perception in *LU* as well. Husserl's analysis of the intuition of the sense object in *PSL* informs that of his account of the sense object *qua* transcendent in *LU*. The *representing* sense material and the object perceived are one and the same.

The Psychical Production of the Transcendent Object

The totalizing psyche functions primarily in its representative mode in accounting for the object perceived as transcendent. The totalizing act may nevertheless be discerned in the process as well. While Husserl clearly invokes the representing mode of consciousness refined in *PSL* to account for the object in *LU*, it is a mistake to interpret this mode as essentially linguistic. He was undeniably concerned to account for the significance of sensible signs in both *PA* and *PSL*. However, beneath these investigations lay his primordial assumption that there could be an awareness of that which was not actually in consciousness *only* on the basis of that which was. Husserl thus breaks the fetishism of the perceptual object *by accounting for it*, and its transcendence, in terms of what is immanent to consciousness. Preeminent in the latter is the activity of consciousness itself.

Husserl frequently presents his theory of sense objects in *LU* in close association with his representative theory of sensible signs of *PSL*. He compares the act of understanding in which a word, written or spoken, becomes meaningful and the objectifying interpretation of sense data by which an interpretation is perceived.[16] Husserl is ever mindful of the analogy between the two. He emphasizes that the sense-"animated" (*belebt*) expression is constituted of a physical aspect and an act. It is the latter which gives it meaning, Husserl writes, and "possibly intuitive fullness." *Only in virtue of the meaning-imbuing act does this physical phenomenon "refer to objectivity."*[17] Only insofar as it does refer to something does the possibility exist that it will find fulfillment in the object intended. Husserl reiterates the assertion of *PSL* that this sense-animated expression is experienced but is not the object of attention. The animating act means or intends something through it. Its animation by sense is simultaneously its transparency.[18]

There are, on the one hand, "acts of word-appearance," Husserl writes, and on the other "*similar* acts of thing-appearance."[19] He asserts that the appearance of the inkwell sitting before him is due to his experiencing a series of sensations objectified by an interpretive act. This act "makes" (*machen*) *this* object *qua* inkwell appear in perception. The word that refers to it, i.e., "inkwell" or "*Tintenfass*," and which seems to "lay itself over" the perceived object, is constituted in a "similar manner."

In both the case of the word and of the object the act is the motive factor. Husserl emphasizes this by stating that when we speak of the knowledge and classification of the perceived object, it *seems* that we are concerned with the *object. But in "experience (Erlebnis) itself . . . there is no object, only perception, the so and so determinate being-expectant"* (*Zumutesein*). Consequently, the act of recognizing his inkwell *as* his inkwell is "founded on the perceptual *act*." This formula is strikingly reminiscent of the assertion in *PA* that totalities predicated on totalities ultimately boil down to acts upon acts. Here, while there is a perceived object, it is nevertheless an *interpretation*. The fetishism of totalities was broken by recognizing them as tantamount to the totalizing act. In *LU* the fetishism of the sensible *relata* of totalizing acts is finally broken by recognizing them to be interpretations of sense data by interpretive acts.

The contents of the totalities of *BZ/PA* functioned as foundations for the act which totalized them. In *LU*, interpretive acts also require the non-acts of sensations. The latter nevertheless remain *immanent* "sensuous kernels" awaiting interpretive light so as to burst forth into sensible interpretations. Inert elements in consciousness are no less immanent for being inert, as Husserl saw in the case of Brentano. When animated, these sensuous kernels do present an object for perception. But this object is nothing more than an interpretation *of immanent material*.[20]

Husserl makes this clear when he criticizes the error that "identifies the distinction between 'purely immanent' or 'intentional' objects and 'transcendent' objects" with that between signs and signified.[21] "The intentional object of the representation" is, Husserl insists, "the *same* as the actual and, in some instances, the external object. . . ." The interpretation stands for nothing other than itself. It is what is perceived as transcendent, which is to say that it is interpreted as transcendent. *It* is the true object, and if it is as it appears, then it is external to and transcendent of consciousness. Husserl does not interpret the "external object" in exactly this way in this passage, yet the term seems to admit only of this understanding. For "it is absurd," he concludes, to distinguish between the intentional object, i.e., the interpretation, and the actual or external object.

The question whether the interpretation is actually external to consciousness cannot be answered by appeal to some notion of its adequation

The Ensoulment of Sensation: Triumph of the Totalizing Psyche

with a *really* transcendent object. Husserl recognizes this as absurd at best and as entailing an infinite progress at worst.[22] The question can be answered only by determining whether or not the interpretation of the sense data is correct. If it is correct, then the object exists as it appears. If the interpretation is incorrect, then the object does not exist as it appears and, presumably, cannot continue to appear to exist. To say that the object is "merely intentional," Husserl writes, is not to assert that it exists. It exists *only in the intentio*, which is to say that the intention, i.e., "a thus constituted object (as) meant, exists." Husserl does not equate the intentional object or interpretation with the intention. The interpretation is not the interpreting act *per sé*, yet it is, he maintains, *a real (reell) part of this act.*[23]

There is no object apart from the interpreting act, only meaningless sensations. There is an object *qua* interpretation when the sensations are imbued with an objectifying sense. The object could not appear in the absence of either sensations or act but the objectifying impetus comes from the latter. In this sense, the object that appears is a function of this act and the meaning it confers. The direction to an object has *nothing to do*, Husserl insists, *with anything* remaining *outside* the representation of the object, *but has to do "exclusively"* with its *"inner peculiarity."*[24]

The totalizing psyche functions almost entirely in its representative mode in Husserl's theory of the interpretation. Yet there are descriptions evocative of the totalizing act of *BZ/PA*. The contents it embraced could be entirely heterogenous, yet in the grasp of this act they became *one*. The act was itself essentially unitary, and its actual grasp was defined in terms of how many entities it could span before splitting up into many acts.

Turning a box over in his hands, Husserl observes that what he sees are not sensations but "*one and the same* box." He infers from changing perspectives that the experienced sensations are themselves heterogenous, and that *the perceived unity of the interpretation must therefore be a function of the interpreting act*. "*Truly different* contents are given on both sides" (by which he must mean *qua* immanent and *qua* interpreted as transcendent), but they are "*meant in the 'same sense.'* "[25]

Sensations are heterogenous and present a unitary perception only having been themselves endowed with the unitary meaning. Here the incorrigibly unitary totalizing act, and the austerely unified meanings intersect, coincide, and complement one another. Here totalizing consciousness *qua* representing and Ideas coincide. Husserl writes that the "actual being or non-being of the object is irrelevant for the true essence of the experience of perception. . . ." It is difficult to determine whether the irrelevance of actual being is due to the fact that perceiving is the conferring of ideal meanings, or to the fact that the latter are conferred by acts which yield only an interpretation — the correctness of which cannot be determined. The ideal implies nothing about reality, the representing-interpreting act produces an interpretation of immanent sense data as transcendent that may be incorrect.

In one stroke, Husserl seems to have accounted for the object perceived as transcendent and to have done so *in terms of the act*. Sensible objects no longer simply stand over against the totalizing act as *relata*. They no longer simply present themselves for analysis by the totalizing intention, or for actual intuition by it. They stand, rather, as *interpretations* generated by it. Their assumed independent existence of consciousness has been decisively revealed for the fetish it is.

What is more, the object of perception, in being accounted for by consciousness, has been relegated to the external sphere where it, *and the probability associated with it*, belongs. Husserl thus cleanses consciousness of the "delusiveness" dragged in by the empirical prejudice. The immanent sphere of acts and sensations remains that in which an *actual* or adequate grasp may be had. Husserl unequivocally espouses the Cartesian position that an adequate inner perception is evident and beyond doubting.[26]

But Husserl's assertion that there may be inner perception of sensations and their animating acts is extremely dubious.

THE DILEMMA: THE UNCERTAINTY OF THE TRANSCENDENT AND THE IMPERCEPTIBILITY OF THE IMMANENT

Husserl sought to account in *LU* for the transcendence of the object. Given his primordial presuppositions, consciousness could account for that which transcended it only on the basis of that which was within and immanent to it. Since in *PSL* he had refined the representative mechanism by which consciousness was preoccupied with that not actually grasped, it was virtually inevitable that the representing mode should be invoked to account, somehow, for the object perceived as transcendent.

But an immanent vehicle for the representing act was required. The perceived object could not be this medium because it was given *as transcendent*. Something immanent must be identified through which consciousness might be preoccupied with the transcendent object, in virtue of which it might perceive. Arguably, Husserl's invocation of sensations as potential *animata* was *required* by his representative presupposition concerning transcendence. These sensations were not givens, they were not what was perceived, they were merely *theoretical posits*. Nevertheless, insofar as they were posited by Husserl *as immanent*, they must be susceptible of inner perception, of adequate grasp. Their presence in consciousness must be indubitable.

But theoretical posits are not perceptible, at least in the usual sense of the word. If Husserl could not grasp these posits in inner perception, and if there was an indeterminacy associated with the correctness of their interpretation, then he was in an epistemological quagmire. Arguably, it was the same one which he excoriated Brentano for falling into.

If that which is transcendent of consciousness cannot be determined to be as it appears, then there can be no certainty that what appears *is* as it appears or that it exists at all. There can be no certainty that the interpretation is correct. Husserl characteristically restricts that of which there may be certainty to what is actually grasped by consciousness, that which is immanent to it, i.e., the interpretive acts and sensations. This realm is eminently accessible to reflection, he believes, although one must be privy to the insight that objects of perception *are* but interpretations in order to engage in "real (*reell*) phenomenological inspection" of it.[27]

Husserl concedes that this inspection is *ex post facto* and not an immediate grasping. Yet he does not seem to entertain seriously the possibility that these contents are then subject to the doubt infecting all memory. Neither does he clarify whether sensations inspected in this reflection are perceived apart from their interpretation or whether interpretive acts are perceived apart from sensations.

If all that Husserl maintains is that the interpretation perceived as transcendent at one moment may be reflected upon as a sort of immanent object deprived of objectification at the next, *then this reflection has not gained truly immanent contents*. It has not because, by his own definitions, the interpretation is not something existing immanently and then interpreted as existing externally. What is immanent are non-interpreted or meaningless sensations and interpreting or meaning-bestowing acts. *If the interpretation is treated as immanent*, even in reflection, then Husserl has sunk into the same quagmire he so vehemently warns others against.

For the "real (*reell*) phenomenological inspection, objectivity itself is nothing," Husserl insists, because it "generally transcends the act." The object is intentional, which only means that "*there is an act there* with a determinately characterized intention." *In* this determination "*is located*," Husserl writes, "what we call the intention toward this object."[28] To say that there is an object perceived is but to say that there are immanent activities which give rise to it. The interpretation cannot be immanent. But since it may be incorrect, certainty can be located only in that which is immanent. Husserl writes that it is the perception directed *exclusively* to "sensational contents" that is "adequate" because these contents "mean nothing other" than themselves. The perception of the interpretation is "inadequate" because the immanent content then "represents that which does not lie in itself, or not entirely so." What it represents is "entirely or partially analogous to it," but immanent content and interpreted object never coincide. If they did, Husserl would be propounding the same mistake which became the empiricistic prejudice.[29] He states that inner perception also "objectifies" that to which it turns, but this does not mean that it is the interpretation that is found in consciousness.[30]

Husserl explicitly proscribes the inference that the interpretation of sensations is a conscious activity. One does not turn toward sensations, interpret them, and then perceive the object produced. These sensations may however become objects of "psychological reflection" which he contrasts to "naive intuitive representation." Ordinary perception remains naive. Psychological reflection is sophisticated in virtue of its knowledge that certainty is found only in the immanent sphere. It can only be done by those "well-practiced" in the "habit of reflection and reflective investigation (that is) contrary (to what is) natural." What is natural is the "naive acceptance" of interpretations as being simply what they appear to be. This acceptance indicates a lack of awareness of the need for epistemological *constraint to the "essential content of acts."* [31]

Husserl allows that we may "simulate" a "consciousness prior to all experiences" (*Erfahrungen*). This consciousness would possess sensations but no objects. Its sensations would receive no objectifying interpretations, they would indicate nothing beyond themselves.[32] Such a "consciousness" would hardly be worthy of the name, Husserl observes, for no one would define it as a "psychical being." It would be a "nonconscious" being or mere body.[33]

It seems unlikely that the psychological reflection Husserl advocates would find such non-consciousness in consciousness. It cannot then be an object of inner perception but only of simulation or fabrication. But if what is perceived is a meaningful sensual configuration, then sensations are neither perceived apart from their interpretation, nor interpretive acts apart from sensations. *What Husserl holds to be actually immanent cannot be grasped as such*. All that may presumably be perceived is the interpretation abstracted from the taking of it as transcendent. That of which certainty may be had is an *immanent object* (if the object of a perception past can be grasped with certainty in reflection). It is not at all clear that Husserl avoided the empiricistic quagmire, however much he was aware of it.

What is more, that which is immanent is the ground for evidential assertions. But what Husserl designates as immanent are insusceptible of the inner perception appropriate to immanent data. Since the immanent sphere is the province of descriptive psychology, Husserl has departed from his own method. Yet the descriptive psychological method or phenomenology which Husserl practiced in the period preceding and including *LU* was tantamount to charting the region of immanent consciousness. The latter was, for Husserl, a totalizing psyche from *BZ* onward.

Husserl could not account for the transcendent object of perception in any way other than the representative mode. The consequence was that his theory yielded neither a transcendent object for perception nor immanent data for phenomenological inspection. At most the theory yielded an immanent object which was the last thing Husserl wanted.

This dilemma would ultimately be resolved by Husserl's embracing transcendental phenomenology. The latter was the only possible resolution *within* the framework of his current philosophy.[34] *Presuming* that framework, he had drawn the fangs of spatio-temporal particularity and the threat it posed to rational norms. The realm of Ideas had been secured, and spatio-temporal reality was but an *interpretation* of immanent elements.

THE GREAT REVERSAL: THE CAUSAL WORLD AS INTERPRETATION

The story of the totalizing psyche from *BZ* to *LU* is in part that of the progressive working out of its intent to totalize contents. Finally, in *LU*, the latter are reduced to inchoate sensations, as meaningless apart from meaning-imbuing acts as the potential *relata* were disparate prior to their totalization in *BZ*. Arguably, Husserl failed in *LU* to give a meaningful account of meaningless sensations. They and their animating acts seem analogous to theoretical posits in physics called for by given observations.

Yet Husserl invokes sensations, even if they may be insusceptible of inner perception. Do not these sensuous kernels, as he calls them, present problems for any interpretation of his scheme as essentially idealistic? Does not their meaning in any case *imply* something external to consciousness which produces them?

It is clear that Husserl does not believe that transcendent or existent objects of any sort can be inferred or deduced from immanent sensational material. According to his theory, sensations in and of themselves refer to nothing. They are inert, inchoate, meaningless givens. Even when imbued with meaning, *sensations* are not intended as transcendent. They remain immanent and are therefore only experienced. It is the interpretation emergent from their animation by the intentional act which is intended or perceived as transcendent. Sensations are therefore taken as indicative of externality neither before nor after their animation by sense.

In his doctrine of sensible objects as interpretations Husserl has effected an inversion of the usual understanding of the sensible world no less radical than that effected with regard to concepts abstracted from particulars. The spatio-temporal world of natural causality is emphatically not the abstractive base for concepts or meanings. The latter do not dwell in reality as a matter of course, but *qua* Ideatic they may confer meaning upon and through it. This natural world is *meaningless* apart from Ideas, apart from the psyche which imbues its sensational stuff with ideal meaning. In the same way that meaning cannot be abstracted from particulars, immanent sensations cannot be regarded as the *products* of external impingements on the body. Both these impingements and the body are putative, they cannot be naïvely accepted as transcendent of consciousness. They are themselves *interpretations of immanent sensations*. Far from being produced by some collusion of the transcendent body and physical entities, the sensations are the stuff out of which all transcendent entities arise *qua interpretations*.

Idealities imply nothing about actual existence. And the interpretations issuing from their infusion into sensational material, also eminently nonimplicative of anything beyond itself, cannot be determined to exist as they appear. Sensational material *qua* immanent is not real but "*reell*," and Ideas are the antitheses to reality. The collaboration of the totalizing psyche and the ideal could not but decisively reverse the supposed dependence of consciousness on the real.

Spatiality is also "*apparent*," Husserl insists, it is also a function of an "objective apperception" of sensation which constitutes it.[35] This is no less true of the causality characterizing spatio-temporal reality — it also is but an interpretation. Causality is emphatically not something found in the immanent sphere by inner perception. As apparent it is subject only to external perception. It is not an "actual given," Husserl insists, and to it corresponds no adequate intuition.[36] The sphere of consciousness remains teleological as in *BZ/PA*. Causality emerges from teleology. To speak of spatiality and causality is to speak of interpretive acts and sensations, of interpretations the correctness of which is indeterminable (or, if not indeterminable, at least dubitable). It would therefore be naïve at best to consider sensations as products of actually existing entities in objective space-time, at worst it would be absurd.

Husserl is nothing if not consistent, and he draws the appropriate conclusions from this theory concerning human bodies and their sensations. "We *appear* to ourselves as members of the phenomenal world." As embodied, we are *interpretations* of sense data within our own consciousness. Other "physical and psychical things (bodies and persons)" appear in like manner, "in physical and psychical relations to our phenomenal self (*Ich*)." Husserl emphasizes that these relations among "phenomena," among interpretations, must be kept separate from those obtaining between elements strictly immanent which found them.[37]

If one's own body is as much an interpretation as any other sensible

The Ensoulment of Sensation: Triumph of the Totalizing Psyche

object, then so are its sense organs. Husserl writes that a "pure phenomenological analysis" of sound finds *only sensations, not "organs of hearing."*[38] The same is true with regard to other organs of sense. There are "sense qualities," Husserl allows, and these are given *whether there are* such things as "senses and sense organs given or not."[39]

The implications of Husserl's doctrine are seen most clearly in his treatment of pain. Pain sensations are really interpretations upon interpretations, or interpretive acts upon interpretive acts. First, given sensations are interpreted *as* some "bodily (*Leib*) member." A sensation of burning pain is then *located* in this member (*qua* interpretation), as well as related to the object taken as the source of the burn. But the sensation of pain is also a "non-act" which must be *interpreted* objectively, as located in the interpretation of the bodily member.[40] Husserl cites other examples including the sadness ascribed to external events, the ache interpreted as residing "in the heart," anxiety tightening the throat, and pain in a throbbing molar.[41] "I take these things precisely in the same true sense," Husserl observes, as I do the wind blowing through the trees. Pain is ascribed to the burned member of the body. This is to say that it is no less an immanent psychical sensation which is *interpreted*, perhaps incorrectly, as spatial or local. *As* interpreted, Husserl insists that these sensations are not evidently given.[42]

In retrospect, this position can be seen to have been adumbrated in *BZ/PA*. There the totalizing act was *external* to its contents as a matter of principle. And this was the case *a fortiori* with regard to sensible contents. Husserl vehemently attacked those who implicitly or explicitly mixed what was psychical with what was local and spatial. The two are kept rigorously apart in *LU* as well. The major difference is that Husserl has succeeded in reducing the spatio-temporal to an interpretation. The threat now is simply that of forgetting that it is an interpretation.[43]

In this respect Husserl's epistemology of *LU* mirrors that of what might be termed "Galilean-Cartesian" ontology. The latter ascribes the origin of the sensible world to imperceptible impingements upon consciousness of external entities. The phenomenal play is considered wholly subjective, it is only erroneously taken as indicative of external reality.

Husserl's doctrine of sense objects accounts for phenomenal perception in terms of similar, although immanent, events. Sensational material is also radically immanent in his view. It becomes meaningful only through the conferment of ideal meanings by immanent acts. These immanent events are also imperceptible, all that is given is their product, the perceptual object. The existence of the latter, *qua* interpretation or product, is always dubious.

The Galilean-Cartesian regard of the perceptual world, and ultimately consciousness itself, as but an interpretation of the true reality according to mathematical physics, is mirrored in Husserl's doctrine which treats external reality as an interpretation of consciousness. Husserl's doctrine is less a deconstruction of the former doctrine than its logical conclusion.

NOTES

1. The analogy to Marx's criticism is drawn by Donn Welton in his article, "Structure and Genesis in Husserl's Phenomenology," found in *Husserl: Expositions and Appraisals* (Notre Dame: University of Notre Dame Press, 1977) p. 54.
2. Edmund Husserl, *Husserliana*, Band XIX/1 and Band XIX/2, *Logische Untersuchungen*, edited by Ursula Panzer. (The Hague: Martinus Nijhoff Publishers, 1984), II, pp. 127—138. All references are made to the first edition of 1901, and give the number of the investigation cited in Roman numerals.
3. *Ibid.*, V, p. 269.
4. *Ibid.*, VI, p. 505.
5. *Ibid.*, II, pp. 127—128.
6. *Ibid.*, II, p. 159. That which is immanent is termed "*reell*" by Husserl in order to distinguish it from that which is not, and, hence, termed "*real*."
7. *Ibid.*, II, p. 160.
8. *Ibid.*, V, p. 469.
9. *Ibid.*, II, p. 193.
10. *Ibid.*, II, p. 159.
11. *Ibid.*, V. p. 327.
12. *Ibid.*, Appendix, pp. 713—714.
13. *Ibid.*, see Appendix, p. 708, for Husserl's definition of sensations as non-acts.
14. This question will be pursued below in the text.
15. *Ibid.*, Appendix, pp. 709—10. Husserl also attacks Brentano's essentially wrong employment of the term "perception" in his doctrine of "inner perception." See V, pp. 350—2 for criticisms of other Brentanonian terms. Husserl is sharply critical here of talk of objects "entering" consciousness.
16. See, e.g., I, p. 74.
17. See I, p. 37.
18. *Ibid.*, I, pp. 30—40.
19. *Ibid.*, VI, pp. 496—7. The interpretation is not the interpretative act, but it cannot arise without it. The expression is given, Husserl writes, in the same way as physical objects, i.e., there is a "certain act-experience in which such and such sensational experiences are 'apperceived' in a certain manner." V, pp. 382—383.
20. The question of whether these sensuous kernels may be understood as indicative of external events, of which they are the products, will be pursued below in the text.
21. *Ibid.*, V, p. 398.
22. See V, p. 408 where Husserl states that the direction toward an object is not to be understood literally in the sense of grasping it like a hand grasps a pen.
23. *Ibid.*, V, p. 399.
24. *Ibid.*, V, p. 408.
25. *Ibid.*, V, pp. 361—2.
26. *Ibid.*, Appendix, pp. 711—12.
27. *Ibid.*, V, p. 387.

28. *Ibid.*, V, pp. 387—88.
29. *Ibid.*, Appendix, p. 711.
30. *Ibid.*, V, p. 385.
31. *Ibid.*, Introduction, p. 12. In *PSL* Husserl ultimately deferred to the "ordinary" view that the object is intuited and rejected the "thing itself" of the psychologists. In that article ordinary perception happened to be an ally of Husserl's, and of totalizing consciousness. Here it is and remains naïve in its obliviousness to the constitutive power of consciousness.
32. *Ibid.*, I, pp. 74—75.
33. *Ibid.*, V, pp. 345—6.
34. The interpretation was given but not certain, the sensations and interpretive acts were supposedly certain *qua* immanent but not given. It is not difficult to see the intentional object of which Husserl already speaks, the *cogitatum*, or the *noema* of which Kern maintains he did not speak until 1904, emerging as the resolution of this dilemma. The interpretation is ultimately granted a berth in consciousness. One may be certain that it is what is being perceived, even if one is not certain that it *is* as it is perceived. It continues to serve only as a clue or index of the *noeses* constitutive of it. Transcendental phenomenology arises as a resolution of this theoretical dilemma beneath which lies not merely Husserl's representational totalizing act, but modern Galilean-Cartesian ontology (which Husserl rejected in later works).
35. *Op. cit.*, III, p. 241.
36. *Ibid.*, V, p. 369.
37. *Ibid.*, V, pp. 328—29.
38. *Ibid.*, V, p. 375.
39. *Ibid.*, Appendix, p. 698.
40. *Ibid.*, V, p. 370.
41. *Ibid.*, V, p. 372; Appendix, p. 704.
42. *Ibid.*, Appendix, pp. 704—711.
43. The preceding account of Husserl's epistemology differs from that, e.g., of Dallas Willard. The latter speaks of an "actual breakout" from consciousness by Husserl through the intentional account of the perceptual object (*LOK*, p. 244). Husserl's primary concern was to *account for* the object given as transcendent *in terms of* totalizing consciousness. His view of the object as "transcendent" and as "existent" must be understood in terms of its being an *interpretation* as such of immanent sense data by immanent acts.

    The issue of the constitution of categories has not been taken up because Husserl insists that such acts are *founded* upon objectifying acts (see, e.g., VI, p. 566). Husserl conceives the objective form of the collection as the realization of a species, and the objective categories associated with individual sense objects as externalizations of psychic connections associated with the constitution of the sense object (see, e.g., VI, pp. 643—44 and 646—50).

    The preceding account differs from that of de Boer also, who maintains that the doctrine of the intuition of Ideas followed rather than preceded that of categorial perception. It seems more plausible that Husserl's Ideatic turn motivated the recognition that the categorial forms of collections could no longer be acts, but must be realizations of Ideas (see *DHT*, p. 154). Neither did categorial constitution motivate the doctrine of sense objects as interpretations, as Fink maintains (see his article, "Operative Concepts in Husserl's Phenomenology" in *Apriori and World: European Contributions to Husserlian Phenomenology*, edited by McKenna, Harlan, and Winters [The Hague: Nijhoff, 1981], pp. 56—70). Rather, it was the natural offspring of the representing act *qua* interpreting (realizing ideal meanings relative to sense material).

AFTERWORD

A Hypothetical Answer for Alfred Schutz

Husserl understood himself in the *Logical Investigations* to have done no less than to have brought the empirical tradition to its logical conclusion and fulfillment: constitutive consciousness. His penetrating critique of the immanent object revealed it to be a fetish obscuring the labors of consciousness producing it. In Husserl's view, the failure of empiricism, until *LU*, lay in its inability to account for the transcendent given *in terms of the activity* of consciousness.

The totalizing act is the transformative nexus between the empiricism mediated to Husserl by Brentano, and the constituting consciousness which Husserl would progressively elaborate. This act coincided perfectly with Brentano's recurrence to psychical acts as loci of apodictic certainty. This recourse to immanent consciousness was required by Brentano's conviction that certainty was to be located neither in the "false" world given in perception, nor in the imperceptible causal machinations productive of it. His faith in unseen causal reality was consistent, in his view, with his agnosticism concerning its specific nature and, hence, with his view of physical concepts as incorrigibly hypothetical.[1]

Alfred Schutz despaired of the possibility of elaborating a satisfactory theory of intersubjectivity within the transcendental parameters of Husserl's later philosophy.[2] However it may be that such an account is not only impossible, but unnecessary. Husserl himself may well have opened the way for its non-necessity.

In later investigations of the origin of geometry, Husserl virtually deconstructed the putative causal world of theoretical physics productive of sensible "appearances." He traced the almost archetypal hypostatization of geometrical idealities which emerged from technical practices such as surveying.[3] Galileo is, in Husserl's view, the paradigmatic modern proponent of this hypostatization. His "hypothesis of hypotheses," that the "objects" of mathematical physics constitute a true reality producing sensible "appearances" both adumbrating and hiding it, entails the relegation of these appearances to the sphere of "subjective" consciousness. Husserl regards Galileo as the modern originator of the model of consciousness elaborated by Descartes and the British empiricists, a model which is a function of his "hypothesis of hypotheses."

It was essentially this model of consciousness propounded by Brentano and Stumpf, and within the context of which the young Husserl began to philosophize. If the older Husserl's analysis of the origin of geometry is correct, including his theory of its hypostatization, and if the totalizing act is compatible with this model and produces constitutive consciousness from it, then it seems that Husserl himself laid the groundwork by which his constitutive assumptions might themselves be dismantled.

If this hypothesis can be confirmed, then the way is opened for a reconsideration of the role of description in philosophy. If it is confirmed, it is clearly the case that the "pure description" practiced by Brentano and Stumpf, and by the young Husserl, was but an alloy of descriptively unjustified presuppositions. It was founded on the hypothesis of hypotheses, on an existent world, as Brentano said, that does not appear, and on an apparent world that does not exist. Since empirical consciousness is a direct function of Galileo's hypothesis, it is but an epiphenomenon of a fallacious hypostatization. Husserl's immanent meaningless sense data and their animating acts are no less theoretical posits. They give rise to interpretations upon interpretations, none of which can be determined to be correct, and therefore to transcendental phenomenology. The *epoché* of the latter reveals only theoretical presuppositions defining the Natural Attitude as an epistemological malady.[4]

Husserl did elaborate descriptive investigations of great penetration and interest which seem incompatible with his presuppositions of constitutive consciousness and the primacy of the transcendental ego. In the second volume of his *Ideen*, e.g., he argues that the other is the means by which the ego is "objectified" and becomes mundane. Yet this objectification presumes the knowledge of oneself as mundane, a knowledge only mediated by the other. In such accounts Husserl suggests a mutually implicative relationship which seems incompatible with the transcendental framework. This is also true of his analyses of touch in which he suggests there is a co-emergence in awareness of the surface of the finger tip and that of the object.

If the transcendental framework can indeed be deconstructed in the way hypothesized, then perhaps a more satisfactory understanding of intersubjectivity can be attained within Husserl's work, an understanding which Schutz himself advanced.

NOTES

1. See earlier points of contact on this issue in, e.g., Descartes' *Principles of Philosophy*, CCIV, and in Locke's *Essay*, Book III, Chapter IV, Paragraph 10, and Book IV, Chapter III, Paragraphs 11—13. One need not search long for other specific points of contact. See, e.g., Locke's distinction between mere sensing and noticing (Book II, Chapter IX, Paragraph 4), and Descartes on "material falsity" (*Meditations on First Philosophy*, Meditation III), and on the fallacious ascription of pain to bodily members (*Principles of Philosophy*, XLVI and CXCVI).
2. Alfred Schutz, *Collected Papers III: Studies in Phenomenological Psychology* (The Hague: Nijhoff, Phaenomenologica vol. 22, 1970), pp. 51—52 f.
3. Husserl regards this hypostatization as virtually complete in the ancient world. He argues that technical practice, e.g., surveying, requires the identification and idealization of various geometrical shapes which, as "ideal limits," place shapes encountered in the perceptual world under a "horizon of infinity." In relation to idealities, perceptual shapes become mere "approximations," they are always "more or less" congruent with the ideal. Husserl argues that this bifurcation of the ideal and real was identified, further, with *epistēmē* and *doxa*, and their correlates of "Being-as-it-is-in-itself," the immutable, self-identical, absolute, and the changing, indeterminate, and relative. The latter was treated as "appearance" only of that which does not change. See his *The Crisis of European Sciences and Transcendental Phenomenology*, trans. David Carr (Evanston: Northwestern University Press, 1970), Part II, and his essay of 1936 in the same volume, "The Origin of Geometry as Intentional-Historical Problem."
4. It is then not quite true that there is "no clear sense to be given to a total suspension of belief in the entire world," as J. N. Findlay writes in his Translator's Introduction (p. 10) to his translation of Husserl's *Logical Investigations*. There is a sense to be given to the *epochē*, but one must be privy to the theoretical presuppositions giving rise to it to grasp it.

Appendix I

The terms, "act of counting" (or enumerating) and "totalizing act" seem to refer to the same psychical act in this introductory chapter. They are not, however, in Husserl's view, the same act. The totalizing act is much more fundamental than counting in that it is the abstractive basis for the number concepts. The latter, *qua abstracta*, are then "attached" to given entities in acts of counting or enumerating.

The act of counting and that of totalizing are not rigorously distinguished in the Introduction for two reasons. They are not, first, in order to emphasize the likely origin of the totalizing act in the lectures of Weierstrass concerning the genesis of the concept of number in the act of counting or enumerating (*Zahlen*). Second, since the full presentation of Husserl's characterization of the totalizing act could not be given until those chapters subsequent to the Introduction, it seemed best to refrain from employing the distinction between counting and totalizing in the latter. The reader should thus come to see both Husserl's continuity with Weierstrass (in the Introduction), and his deepening of this inherited notion by distinguishing it from simple counting acts (in subsequent chapters).

The term, "totalizing act," is, then, neither a translation of "*Zahlen*" nor of any other German term of Husserl's. The author came to employ this term for the following reasons.

In his *Habilitationsschrift* of 1887, Husserl writes:

". . . in all cases where discrete contents are thought-together, i.e., in a totality (*Inbegriff*), there is extant (*vorhanden*) one and the same, constantly uniform act of embracing (*zusammenfassen*) interest and noticing, which separates the individual contents each for itself (i.e., as noticed for itself) and holds it together unified, simultaneously, with the others."

(see p. 337, 24—32, of *Philosophie der Arithmetik, Husserliana*, Band XII, ed., Lothar Eley [The Hague: Martinus Nijhoff Publishers]). The basis for the emergence of the term, "totalizing act," is found in the author's translation of *"Inbegriff"* as "totality." Now this word may also be translated as "aggregate." In fact, Cairns writes that " 'aggregate' may be the best trln." However, just prior to reaching this conclusion he writes, "Save 'aggregate' . . . for '*Aggregat*.' " Clearly Cairns is somewhat conflicted as to how it should be translated. And if he says that "aggregate" should be saved for its German cognate, but then encourages its use as a rendering of *"Inbegriff,"* we might, similarly, acknowledge that, as Cairns says, "totality" should be saved to translate *"Allheit,"* but nevertheless employ it as a rendering of *"Inbegriff"* (see Dorion Cairns *Guide for Translating Husserl* [Martinus Nijhoff, The Hague], Phaenomenologica, Vol. 55).

Another reason that *"Inbegriff"* was not translated as "aggregate" was that Husserl mentions both *"Aggregat"* and *"Inbegriff"* in the same breath as terms that are more or less equivalent but which are "not without appreciable nuances" which serve to distinguish them (see his *Habilitationsschrift* at p. 297, 31—36, of the above-mentioned *Husserliana* volume, *Philosophie der Arithmetik*). If "aggregate" is used for *"Aggregat,"* as it clearly should be here, then another rendering than "aggregate" must be found for *"Inbegriff."* The author concurs with Professor Willard at this point that "totality" is the most suitable rendering (see his essay, p. 96, in *Husserl: Shorter Works*, McCormick and Elliston, eds., University of Notre Dame Press, Notre Dame, Indiana, 1981).

Husserl thus distinguishes *"Aggregat"* and *"Inbegriff."* According to his notes, Weierstrass *did* employ the term, *"Aggregat"* for the collection of entities generated by the "operation of counting" or "enumerating" (*Zahlen*), entities which shared some "relevant characteristic" as the basis for their selection (see p. xxiv, note 1, of the *"Einleitung des Herausgebers"* in aforementioned *Husserliana* volume, *Philosophie der Arithmetik*). This emphasis on "things" possessing some common characteristic is certainly in keeping with the rootage of *"Aggregat"* in *"grex."*

Husserl clearly rejected the stipulation that the things collected must be similar in some respect. While the act described in the quotation above may unite similar entities, it is its capacity to embrace radically *dissimilar* things which distinguishes it, as will become clear in subsequent chapters. It thus seems likely that Husserl, in departing from what he understood to be Weierstrass's view on this matter, also departed from his characterization of the collection as an *"Aggregat."* Thus, it seemed to the author that to translate *"Inbegriff"* as "aggregate" failed to do justice to this departure, as well as to Husserl's aforementioned distinction of *"Aggregat"* and *"Inbegriff."*

Given these reasons for translating *"Inbegriff"* as "totality," it seemed quite reasonable to the author to *characterize* the psychical act of "unified interest and ... noticing," this act which "lifts out and embraces different contents for themselves," thus unifying them *qua* "totality," as the totalizing act (see the *Habilitationsschrift* at p. 333, 28—35, in the *Husserliana* edition, *Philosophie der Arithmetik*). The term is a *characterization* of this act because, to the knowledge of the author, Husserl does not himself refer thus to this act — yet it seems virtually present in that he speaks of the act which (as will be demonstrated in Chapters 1 and 2) is *tantamount* to the totality.

It should be emphasized as well that the term, "totalizing act," is intended by the author to be neither a characterization nor a translation of Husserl's term, "*die kollektive Verbindung*" (translated in this work as "collective connection") or of the verb, "*kolligieren*" (translated as "colligating [act]") which Husserl correlates with the emergence of a "*Kollektion*" (see his *Habilitationsschrift*, p. 305, 6—10, of the *Husserliana* volume, *Philosophie der Arithmetik*).

As will be demonstrated in subsequent chapters, the "totalizing act" must be distinguished from the "collective connection." The latter, Husserl writes, can "be noticed ... *only by means of reflection on* the psychical act ... ," i.e. the totalizing act (see his *Habilitationsschrift* at p. 333, 32—35, or his *Philosophie der Arithmetik* at p. 74, 10—13, both in the above-cited *Husserliana* volume XII).

"*Verbindung*" was translated as "connection" because, while both "*Verbindung*" and its natural rendering, "combination," exhibit the Latin, "*bini*," Husserl's examples are almost always of *more* than two entities held together in thought. In the author's view, then, "connection," since it does not root in "*bini*," better conveyed what was arguably Husserl's intent and what was without a doubt his practice than does the term, "combination."

Appendix II

Willard asserts (*LOK*, p. 42) that *BZ* 332, 6—16/*PA* 72, 23—29 conforms to *BZ* 301, 1—21/*PA* 20, 8—28 (*LOK*, p. 41) in that it stresses the "observability of the unification in the concrete totality." He states that the passage is "word-for-word the same" in each version, but fails to note that *BZ* 332, 12—16 is omitted in *PA* (even though he makes his case primarily on the basis of the latter text, and only secondarily on that of *BZ*). He proceeds to argue, based on a *PA* interpolation of the *BZ* text, in a vein he believes reinforces his point here concerning the contentual status of the "unification."

For further discussion of his view of the collective connection as "objective" *qua* contentual, as "found uniting the members of any concretely given multiplicity" (*LOK*, p. 62 — the footnoted expansion of this point that Willard provides [*LOK*, p. 84, n. 76] provides no textual justification, and that which is intended as such is arguably anachronistic, since based on the *Logical Investigations*) see especially pp. 54, 60, and 62.

Willard leaves little question remaining concerning his interpretation of *BZ* when he writes that, "all methodological subtleties aside ... this essay ... advances an eidetic and noematic analysis...." (*HSW*, p. 118, n. b.). Like many interpreters of *BZ/PA*, he tends to treat the collective connection, the totality, or some combination of the two, as a categorial noematic object (cf., his statement in *HSW*, p. 91, n. 3, where he also cites the passage under consideration). In the introduction to his translation of *BZ* he quotes Farber's statement approvingly, "that Husserl means to name something objective when he speaks of totalities or pluralities." To reinforce this objectivist interpretation, Willard cites the "categorial" concepts which apply to everything, and asserts that these are also evidence of Husserl's "non-subjectivist analysis."

While Husserl's analysis is undoubtedly non-subjectivist, the latter is not convertible with the "objectivist" position on totalities. The latter can be denied while the former is maintained.

Appendix II 133

While J. Philip Miller pays close attention to de Boer's position on this matter (de Boer bases his position, which seems the correct one, mainly on texts from *PA* not found in *BZ*; it will be considered in Appendix III) but he is, nevertheless, in basic concurrence with Willard that the general concept of multitude is to be abstracted from concrete multitudes (see his *Numbers in Presence and Absence: A Study of Husserl's Philosophy of Mathematics* [The Hague: Martinus Nijhoff, Phaenomenologica, vol. 9, 1982], p. 40 [hereafter *NPA*]). Miller appeals also to an interpolation of *PA* (*NPA*, p. 40) in the *BZ* text which will also be considered in Appendix III.

Miller contends (*NPA*, p. 38) that Husserl uses "the verb '*entstehen*' at a pivotal point (one already cited by the present work, i.e., *BZ* 333, 28/ *PA* 74, 7) . . . with reference to a concrete object rather than a concept. A concrete aggregate 'comes to be' (*entstehen*). . . ." It is evident that Miller assumes that because the totality is concrete, it is also objective. If this is not his precise line of reasoning, it is nevertheless clear that he takes the totality to be contentual in the sense of being "more than" or "beyond" the embracing act itself. Yet the context of this quotation seems to militate against precisely such an interpretation.

Miller appears to make the same mistake in the next paragraph that Willard (and Farber) make. He states correctly that concepts ("the aggregate itself," still *NPA*, p. 38) are distinguished from the particular acts which intend them (this seems to be the "non-subjectivism" found also in *LOK*). He then states that a "concrete aggregate is not identical with the presenting act. . . ." This seems to imply that the totality is in some manner contentual (the objectivist thesis of Willard [Farber]). In the latter part of this sentence Miller cites a passage (i.e., one also considered already in this work, *BZ* 309, 9—19/*PA* 31, 10—21) and asserts that it states that "a concrete aggregate" is the "meaning" (*Bedeutung*) or "logical content" of "the presenting act." It is not entirely clear what Miller means here. The point of the passage he cites is that the concrete phenomenon must be distinguished from its "meaning." There is a "presenting act," as Miller writes, of the concrete totality, which is the embracing act itself. There is also a presenting act of the concept of the concrete totality. The two presenting acts are separate. Further, the fact that the act directed toward the concept is not identical with the concept does not entail the non-identity of the totalizing act and the totality of its contents. They are not, technically, identical but equivalent — although Husserl does not seem to make this distinction and thus may well have regarded them as virtually identical. The psychical whole *is* the "thinking-together psychical act."

Sokolowski (who is essentially a proponent of the position criticized, and who will be considered in Appendix III) cites his inclination "to accept the thesis of Walter Biemel, who claims that the problem of constitution is already operative, in rudimentary form, in Husserl's first published work" (see Sokolowski's, *The Formation of Husserl's Concept of Constitution* [The Hague: Martinus Nijhoff, Phaenomenologica, vol. 16, 1970], p. 6, n. 3; hereafter *FHCC*).

Biemel's claim is that the process of abstraction and consequent orientation toward the conceptual is the germ of the noematic object of transcendental phenomenology (see *The Phenomenology of Husserl*, ed., R. O. Elveton [Chicago: Quadrangle Books, 1970], pp. 148—73 for Biemel's paper delivered at the International Phenomenological Colloquium at Royaumont in 1957, "The Decisive Phases in the Development of Husserl's Philosophy;" hereafter *DP*).

Biemel seems to maintain also that the totality is contentual. He cites a sentence of the same passage cited by Miller which states: "The collective connecting can only be grasped by reflecting upon the psychical act through which the aggregate comes into being" (*BZ* 333, 24—334, 19/*PA* 73, 16—74, 20 [*DP*, p. 152]). Following this quotation, Biemel appears to make two mistakes when he asserts that the "psychical act through which a significance, a meaning, comes into being is to be grasped through reflection" (*DP*, p. 152). In point of fact, Husserl writes, "the collective *connection*," not "connecting" (and it is this which is grasped in reflection on the psychical act, not the act itself). Biemel refers to the "totality" in Husserl's sentence as a "significance" and a "meaning" which comes into being in virtue of the psychical act. Biemel's point seems to be that the act brings into being something contentual. If he were referring here to the concept of such, he might be correct. But it is clear that he is referring to the concrete totality.

The heart of Biemel's thesis follows (*DP*, p. 153). He cites the passage in which Husserl insists that numbers are not "productions" which continue to reside in time and space in the absence of the producing act (*BZ* 317, 7—17/*PA* 45, 27—46, 7). Biemel then states, "That there are structures which must be produced in thought in order to exist, and which therefore exist only insofar as they are produced in this way, that is, insofar as determinate thought-processes are set in motion, is *in nuce* the idea of constitution which occurs to Husserl in his attempt to grasp the essence of number."

Biemel is quite correct if he refers here to number concepts, for they do depend wholly on thought-processes. Yet while these concepts are certainly non-subjectivistic, this by no means entails their being contentual, even in some "eidetic" sense. Perhaps the most fundamental critique of this construance of concepts as proto-noematic objects (and Biemel says that we see in *BZ* "the method of exhibiting the essence of an affair by returning to the origin of its meaning in consciousness," *DP*, p. 152) is simply to point out that in the case of the latter, the *noema* is taken as "transcendental clue" to the *noeses* which constitute it (for a discussion of the various meanings of "origin" in Husserl's *corpus*, see de Boer's *DHT*, pp. 71—73). Quite to the contrary here, Husserl emphasizes that concepts are not deposits which may be "found," and which could then conceivably serve as such clues. Biemel might argue that they are found *qua "applicata*," or in the various mathematical "recipes" for their generation, and Husserl is demonstrating simply that such have their origin in thought.

Yet this example does not seem to be sufficiently noematic, or distinctively so, to yield what Biemel would like it to. Husserl's is essentially a classically empiricist doctrine of the abstraction of concepts. If the latter is viewed as a salient antecedent of transcendental phenomenological constitution, then so may be this particular instance of it.

Appendix III

PA 69, 35—70, 1 occurs within the context of an interpolated and expanded (relative to *BZ*) passage in *PA* (*PA* 69, 33—71, 22). The latter is mainly constituted by Husserl's attack on Drobisch's view that, "since every compounded relation necessarily embraces, as a relation-complex, more than two *relata*, conversely, every relation with more than two *relata* is necessarily a complex of relations." Husserl cites the counter-examples of the collective relation and of similarity between sense objects as instances of relations between arbitrarily many *relata* which are simple. He defines compound relations as those compounded from other relations, and simple relations as those which are not.

Some of the passages cited in this chapter are those on which de Boer bases his position that there are no "objects" of higher order acts other than the acts below them, and that, ultimately, it is only by means of first order acts that those above them extend to "primary contents." This is Husserl's position. de Boer's argument does not directly address (as the present work does) the question of whether the simple totality, multiplicity, whole, or even connection, is in some manner contentual.

de Boer cites Husserl (*LU* II, 282, n. 1; *DHT*, 11, n. 5) as (by the *Logical Investigations*) maintaining that *PA* was concerned both with "acts and objects of a higher order." de Boer regards this assertion as true with regard to acts only. He also cites a passage from *LU* (*LU* II, 401 ff.) as implicit self-criticism on Husserl's part, for the latter maintains there that even composite acts have single objective correlates (*DHT*, 26, 14). de Boer seems to assert that Husserl was, in *LU*, both aware of his position in *PA* (i.e., higher order acts have no objects as such) and unaware of it insofar as he refers to "objects of a higher order" (*DHT*, 25, nn. 6—7). He cites Husserl's "Draft of a Preface to the 'Logical Investigations' " of 1913, as a more accurate recollection on Husserl's part of the nature of his position in *PA*.

It is indeed true that Husserl maintains, as de Boer points out (*DHT*, 28; see also 119 for a summary of de Boer's position) that number is not a sensory "part-content" which may be abstracted, and that he attacks those theorists who construe it as such. However, de Boer seems to *identify* its non-sensuousness with its non-contentuality. The former is a case of the latter, but the latter is not exhausted by the former, simply because, as Husserl points out (*PA* 68, 31—79, 8/*BZ* 330, 12—25), the "foundations" or *relata* are either primary/physical or psychical. The latter are obviously not sensible, yet the relation is still contentual in the sense that it is perceptible in some fashion in and among these foundations. This interpretation of de Boer's view seems to be borne out by the references he gives (*PA* 71, 5—72, 1; 73, 16—74, 20/*BZ* 331, 10—26; 333, 10—334, 3), in which it is true, as he says (*DHT*, 28, n. 25), that these passages distinguish content from act, and locate the collective connection in the latter. However, de Boer *seems* (*DHT*, 25) to identify content exclusively with *sense* content. But while contents may be psychical, the (psychical) relations are not intuitable in them. Of this de Boer is himself wholly aware (see *DHT*, 18).

Husserl states explicitly that certain relations (e.g., identity, similarity) hold between psychical acts or states (judgments, acts of will, etc.) and are, therefore, physical or primary (*PA* 68, 31—69, 8/*BZ* 330, 12—25). The problem, if there is one, is not so much Husserl's "sensualism" (*DHT*, 29) as it is his founding of number on psychical *acts* at this point. de Boer's emphasis on the sensuous may reflect an assumption that numbering or counting is directed, primarily, toward sensible objects (the shepherd, e.g., counting his sheep at night as they pass single file through the door into the fold), and only derivatively toward imagined objects (the shepherd "counting sheep" in order to fall asleep). But totalizing is not limited to, nor occurs, even in a primary sense, only at the sensible level. And it is this act on which Husserl bases the derivation of numerical concepts which are then "applied" to whatever — perhaps, primarily, sensible objects.

de Boer's position to the effect that there are no higher order objects, as correlates of higher order acts (other than those acts) in *PA*, unfortunately becomes rather more compromised in his discussion of simple totalities and the abstraction of the collective connection. He seems to neglect, as most commentators do (*DHT*, 30), the mechanism of "attaching" concepts to concrete contents in his assumption of the ambiguity of Husserl's description of the process of the abstraction of the collective connection. The passage de Boer cites in this regard (*PA* 79, 1—27; the contents of this passage are essentially new in *PA*, but parallel those of *BZ* 337, 9—18) is, indeed, concerned with abstraction and, arguably, with abstraction "post-attachment" (with the abstraction of the concept of multiplicity). de Boer writes that, according to Husserl, "on the one hand we reflect on an act, while on the other hand the collected contents must not disappear from consciousness. In a certain sense, then, the contents themselves serve as the basis for the abstraction" (*DHT*, 30, n. 34).

Now Husserl is concerned with abstraction from concrete contents. But the problem is simply the question whether the concrete contents disappear from consciousness when abstraction is taking place and, with them, the particular connection. Husserl contends that *neither* (contents nor connection) do, simply because "to disregard or to abstract from something simply means: to notice nothing particular concerning it." This non-attending neither produces new contents, in the sense of deposits which may be found subsequently, nor alters ontically those to which it is directed. Also, it is not the "act" which is kept in view here, as de Boer states, but the "connection" (which is, technically, not convertible with the totalizing act). Husserl does, indeed, in this passage, use "contents" (*Inhalte*) as what seems to be shorthand for "particulars" (*Besonderheiten*) at points (see, e.g., *PA* 79, 19—24). It is evident, however, that his reference is to particular contents, or the *particularity* of the contents, in contrast to the "conceptual extract." The attention in this passage to parts and their manner of connection is reminiscent of the discussion at *PA* 72, 17—19 (*BZ* 331, 38—332, 3).

While the apprehension of contents and their collection is the "precondition" for abstraction in this passage, this does not imply that such occurs, in any primary sense, from concrete multiplicities (concrete here, as elsewhere, means particular and not necessarily sensuous). In fact, Husserl immediately states that the "lifting-out interest" extends not to the contents, but exclusively to their "thought-connection." The "main interest" concentrates on the "collective connection." This, as Husserl argues elsewhere, motivates the recognition that such connection is none other than the totalizing act, and is not a content at all, which in turn initiates the circle of reflection, abstraction, etc., and, finally, "attachment."

Other scholars who have regarded either the totalizing act, collective connection, totality, or some combination of the three, as contentual, in either *BZ*, *PA*, or both, are Willard, Miller, Biemel, and Sokolowski.

Miller notes that "Husserl allows, of course, that once we have determinate number concepts, we can apply them to concrete multitudes (*PA* 335f.). The point here is only that these concepts do not *arise* in concrete experience according to the doctrine presented in *BdZ*" (*NPA*, 44, n. 41). This does *not* mean that Miller believes the abstractive basis not to be the "multiplicities." On the contrary, he propounds this doctrine with regard to *PA* as well. He cites, in this endeavor, *PA* 82, 2—6 (*BZ* 337, 4—7). This passage, in *BZ*, occurs within the context in which "attachment" is presumed, and is immediately followed by precisely the contrary interpretation to that which Miller wishes to advance (*NPA*, 92).

Miller bases his position, also, on *PA* 81, 31—34a (roughly, *BZ* 336, 39—337, 1). The context of the passage in *BZ* is the "falling under" the concept of the concrete multitude. Hence, at *BZ* 337, 1—2, Husserl observes that it is "easy" to characterize the abstraction of the concept from the latter. Miller cites the "brief though important addition" to the "version presented in *PA*," and quotes the latter: " 'To derive the number concepts it is not necessary of course to take the general and indeterminate concept of multitude as mediator. We come directly to these concepts from any concrete multitude whatever . . .' " (*NPA*, 40). Now it is, indeed, possible that such an interpolation in *PA* is indicative of a shift toward a "categorical *object*." However Miller omits the remainder of the sentence, which is more or less the correlate of *BZ* to which we have alluded: ". . . *because* each one (concrete multitude) falls under one, and truly one, of these determined concepts." Arguably, the phrase "falls under" indicates the prior possession and application of the concept to concrete multitudes, from which it is subsequently "abstracted." Miller also adduces *PA* 79, 6—10, which has been treated within its larger context above.

Sokolowski's contentualism tends to embrace both the collective connection and the totality. He maintains, e.g., that the former is a "*concretum*" (*FHCC*, 16) and cites *PA* 15, 24—16, 20 as evidence. If his reference is to *PA* 16, 11 ("the *concreta*"), then Husserl's reference in this sentence is, as it says, to the "totalities of determinate objects." Sokolowski also cites *PA* 18, 31—37, in this regard. Here, it is the connection that is mentioned, but it is not specified *where* it is to be found.

Sokolowski seems to treat the "group" as something contentual (in his terms, "objective") and by this he seems to refer to something more than either the contents or the totalizing act. He also acknowledges that the group is a function of the latter, and so he is forced to characterize Husserl's description of groups as "paradoxical" (this is not, however, an inconsistency on Husserl's part, but is due to the "paradox" of groups themselves and, indeed, of "all psychic relationships" [*FHCC*, 16—17]). It would seem that Sokolowski has simply failed to appreciate the rigor with which Husserl keeps the psychical totalizing act and the contents separate. He does accuse Husserl of confusion at one point (*FHCC*, 17, n. 2), asserting that he fails to keep distinct the "concrete phenomenon of a psychic relationship, and the abstract concept arising from reflection on the mental act that forms concrete phenomena." Here, again, is the assumption that the "group" is a "concrete phenomenon" that is "formed" by the "mental act" — the model is clearly noetic-noematic. Husserl does indeed speak of the "concrete totality" as the counterpoint of such, but the former is simply the specific act totalizing given contents. Sokolowski also fails to take due notice of the mediating role of "attaching" acts, and the subsequent acts of "abstracting," which render this somewhat less confusing.

He cites, in this regard, *PA* 87, as an apparent example of the correlativity of the act and the contentual relationship established by it. In this passage Husserl is distinguishing two aspects of a doctrine — the abstracting function of the "thought-act" and the function of the latter as *that* "*from which*" the concept is abstracted. In the latter case, there is a correlation between the particular and the concept under which it ranges, but it is not the "correlativity" of *noesis-noema* of the later phenomenology, as Sokolowski seems to suggest (*FHCC*, 17).

He appears to jump, here, from the act-concept relation, just discussed, to what he takes to be the act-relation relations. In the latter case he seems to believe that Husserl thinks that the connection is somehow contentual. This same confusion persists at, e.g., *FHCC*, 22. Sokolowski is quite correct that "subjective mental acts" are responsible for "logical and mathematical categories," but these are concepts. Further, they are not concepts of objects or contents and are not themselves contentual. (Also at *FHCC*, 22, Sokolowski maintains that the number concept gained *via* reflection on the psychic act is equivalent to the confusion of the *noema* with the *noesis*, thus giving us some insight into his hermeneutic assumptions.) de Boer gives an accurate, but scathing, critique of Sokolowski's views along these lines, as found on *FHCC*, 18. There, according to de Boer, he makes "all the mistakes that one could make in interpreting Husserl's *PA*" (*DHT*, pp. 119—120, n. 9).

Sokolowski is quite correct (*FHCC*, 19) when he maintains that *PA* does "suppose that a subjective process produces a form or category which is no longer subjective, but in some way transcends the subjectivity which produces it." But such logical or conceptual content is of acts. It is objective in the sense of non-"psychological" or particular, but it is not so in the sense of ideal *contents* (see p. 81 f., also, on this matter, with reference there to *BZ*). Sokolowski then argues, correctly, concerning Husserl's anti-psychologism in *PA*, but his statement that, in this anti-psychologism, Husserl fails to "sufficiently distinguish" the "logical categories, and . . . psychological facts" seems manifestly unfounded (*FHCC*, 21; see, in this regard, *PA* 31, 14—21; 33—44/*BZ* 309, 13—19; 32—33).

However, at *FHCC*, 23—24, Sokolowski correctly assesses the doctrine of *PA*, when he concedes that "we cannot say that logical and mathematical categories are directly constituted by mental acts," noting that, in *PA*, "category" refers to concepts arising from reflection on mental acts, not to the "relationship" that is "present" in the concrete phenomena of groups produced by act. It is clear, nevertheless, that Sokolowski maintains the interpretation here which has been argued is untenable.

The difficulty, he writes, is "mainly one of terminology," on Husserl's part (and Willard agrees this is the case as well; see, e.g., *LOK*, p. 54, where Husserl's "choice of terminology" is described as a "disaster;" cf., also, p. 60 in this regard, as well as his introduction to the translation of

BZ in *HSW*, p. 88), and then cites *PA* 20, 17—20 (*BZ* 301, 9—12) as evidence of a contentual and "peculiar unification there" which is "present in what he (Husserl) calls the concrete phenomenon." And, Sokolowski concludes, this "peculiar unification" can then "be considered as the logical form constituted by a mental act." As has been argued with reference to this passage's antecedent in *BZ*, an interpretation of this sort is tenuous when erected on the ambiguity of this passage. In any case, Sokolowski's point here seems contrary to his intent in his own passage (*FHCC*, 24).

Gottlob Frege may well have been the progenitor of the interpretations of Husserl's *PA* considered. In his review of *PA*, Frege accuses Husserl of the "paradoxical" description of the totality (to borrow Sokolowski's term). He asserts that "totality (set, multiplicity) appears now as a presentation (pp. 15, 17, 24, 82), now as something objective (pp. 10, 11, 235)." (*HEA*, p. 316). The following are those passages in which Frege alleges that Husserl points to the "objectivity" of groups. At *PA* 15, 24—17, 10, the reference is clearly to "concrete totalities" (*PA* 10—11). Part of p. 10 was contained in *BZ*, the rest is new to *PA*. At *PA* 209, 14—210, 10 (*PA* 235), the discussion is largely centered on figural moments and their symbolic relation, as spatial configurations, to larger groups of contents. This passage is also new to *PA*.

Those passages in which Frege finds a "subjective" construance include *PA* 19, 34—20, 29 (*PA* 15). Most of this passage is also found in *BZ*. It involves the discussion of the extraction of the concept of the relation from the psychical act. At *PA* 22, 1—24 (*PA* 17), Husserl describes the totality of objects as equivalent to their co-presence in consciousness (the correlate of this passage in *BZ* has been discussed). At *PA* 27, 35—28, 14 (*PA* 24), a passage also found in *BZ*, the discussion concerns the psychological preconditions in totalizing contents — as contrasted with the content of the concepts. And, in the last passage mentioned, one also with an antecedent in *BZ*, *PA* 77, 1—23 (*PA* 82), Husserl treats the abstraction of concepts from psychic acts which lead to the concepts of multiplicity, totality, etc.

Frege's review of Husserl's *PA*, and the nature of his influence on Husserl, or lack thereof, has been the subject of much investigation. It seems to be the consensus that his direct influence on Husserl was not monumental. Willard believes that Husserl was influenced much more significantly by the "mathematizing logic" of George Boole and Ernst Schröder (see his *LOK*, pp. 118—124; see also Mohanty's monograph, *Husserl and Frege*, and for an interchange between him and Føllesdal on the subject, see *Husserl, Intentionality, and Cognitive Science*, ed. Hubert L. Dreyfus [Cambridge and London: The MIT Press, 1984], pp. 43—56).

Appendix IV

"*Eigentlich*" has been translated as "actual" and "*uneigentlich*" as "inactual" instead of the more traditional renderings of "authentic" and "inauthentic." In the author's view, the former terms better convey Husserl's sense in English. The reader may yet protest that "actual" should be reserved for translating "*wirklich*," and that, further, the meanings in German of "*eigentlich*" and "*wirklich*" are not convertible.

In practice, however, Husserl does not seem to distinguish them, and, in fact, employs them interchangeably; this is not insignificant in the practice of one capable of discerning the subtlest nuances in meanings of the terms he invoked. At *PA* 192, 11—15, he observes that "only under especially favorable circumstances can we actually (*eigentlich*) represent concrete pluralities of approximately a dozen elements . . ." He proceeds to reiterate and elaborate this by writing, ". . . *that is, in fact (faktisch*), to 'lay hold of' (*befassen*) each of its members as noticed for itself together with all the others." It seems clear that Husserl here understands the "*eigentlich*" representation of the plurality of this number as amplified in meaning by the characterization of it as a "'*faktisch*' laying hold of" the represented members. The latter word would seem to be closer in meaning to "*wirklich*" than to "*eigentlich*," yet Husserl seems to use "*faktisch*" virtually as a synonym for "*eigentlich*."

More significantly, in his discussion of his thesis that "Arithmetic does not operate with the 'actual' (*eigentlich*) number concepts" (*PA* 190, 5—192, 32) Husserl uses, as far as this writer can discern, the words "*wirklich*" and "*eigentlich*" interchangeably. In the title of the section just quoted, he speaks of "*eigentlich*" number concepts, yet at line 13 (*PA* 190) he speaks of "*wirklich*" numbers. Since even the lowest level of the concept remains conceptual, his omission of the word "concept" in conjunction with that of "number" at this point can be read only as shorthand. Husserl is here contrasting those concepts "indirectly symbolized" and those not. The latter derive from totalizing acts entirely practicable while the former indicate those impracticable.

Husserl employs these terms interchangeably again in this section in their modification of "representation" (*Vorstellung*). He writes that "we can ascribe the actual (*eigentlich*) representation of *all* numbers only to an infinite understanding" (*PA* 191, 35—36). He proceeds at *PA* 192, 2—4, to observe that even if an "infinite being" were imaginable, it would require the light years of the astronomers to gain "the actual (*wirklich*) representation of millions and trillions . . ." There is little doubt that the object of reference — this *type* of representation of given numbers — is the same at both points in this passage; Husserl speaks of this representation as "*eigentlich*" at one moment and as "*wirklich*" at the next.

At *PA* 192, 28—30, Husserl contrasts the givenness of "symbolic" (*symbolisch*) and "actual" (*eigentlich*) number concepts, as he does at *PA* 193, 4—6, and elsewhere. As has already been observed, he makes essentially the same comparison using "*wirklich*" instead of "*eigentlich*" (*PA* 190, 8—17).

With reference to houses instead of numbers, Husserl states that we have an "actual (*eigentlich*) representation of the external appearance of a house if we actually (*wirklich*) observed it" (*PA* 193, 13—14). One might protest that Husserl is linking "*eigentlich*" with "*Vorstellung*" only, but his modification of the latter term by "*wirklich*" (at *PA* 192, 2—3) would seem to silence this protest. If one suspected, further, that the latter use was restricted only to representations of number, then one would find that Husserl also speaks of "*wirklich*" representations of "intuitable objects" — e.g., houses and streets, of which he is at this point speaking (*PA* 194, 1—2).

Husserl uses these terms interchangeably as well in the section entitled, "Attempt Toward Clarification of the Momentary Grasp of (Sensible) Multitudes" (insertion mine; *PA* 196, 13). There he writes that "further analyses of a psychical act" are needed in order to clarify the "actual (*wirklich*) representation of a multitude" (*PA* 196, 22—25). At another, and slightly later, point he writes — interchanging "*Menge*" and "*Vielheit*" as well — that "it is indubitable (that) the concrete representation of a plurality is not actual (*eigentlich*) here" (*PA* 197, 12—13).

Based on such instances as these, which are certainly not anomalous, it seems that no clear line can be drawn between "*eigentlich*" and "*wirklich*" as Husserl employs them in these texts. More important, in *his* view, there does not seem to be reason to draw such a line. This does not seem, further, to have been but the temporary absence of such a line. In his article of 1894, "*Psychologische Studien zur elementaren Logik*" (treated in Chapter 5 of this work), Husserl virtually relies on the term "*wirklich*" with reference to representations to which he earlier clearly applied the term, "*eigentlich*." For instance, he contrasts the "actual" (*wirklich*) contents of our representations" with those "merely intended contents" (see p. 167 of this article, *Philosophische Monatschefte* 30, 1894, pp. 159—191).

Bibliography

Adorno, Theodor W. *Against Epistemology: A Metacritique*. Translated by Willis Domingo. Cambridge: The MIT Press, 1983.
Apel, Karl-Otto. *Towards a Transformation of Philosophy*. Translated by Glyn Adey and David Frisby. Boston: Routledge & Kegan Paul Ltd., 1980.
———. *Understanding and Explanation*. Translated by Georgia Warnke. Cambridge: The MIT Press, 1984.
Bachelard, Suzanne. *A Study of Husserl's Logic*. Translated by Lester E. Embree. Evanston: Northwestern University Press, 1968.
Beth, Evert, W. *The Foundations of Mathematics*. Amsterdam: North Holland Publishing Company, 1968.
Boyer, Carl B. *The History of the Calculus and its Conceptual Development*. New York: Dover Publications, Inc., 1949.
Brentano, Franz. *Psychology From an Empirical Standpoint*. Translated by Linda L. McAlister, Antos C. Rancurello, and D. B. Terrell. New York: Humanities Press, 1973.
———. *Sensory and Noetic Consciousness*. Translated by Linda L. McAlister and Margarete Shattle. New York: Routledge & Kegan Paul, 1981.
———. *The True and the Evident*. Translated by Roderick M. Chisholm, Kurt R. Fischer, and Ilse Politzer. London: Routledge & Kegan Paul Ltd., 1966.
———. *The Origin of our Knowledge of Right and Wrong*. Translated by Roderick M. Chisholm and Elizabeth H. Schneewind. London: Routledge & Kegan Paul Ltd., 1969.
Carr, David. *Phenomenology and the Problem of History*. Evanston: Northwestern University Press, 1974.
Chisholm, Roderick, M. *Brentano and Meinong Studies*. Atlantic Highlands: Humanities Press, 1982.
Christensen, Darrel, Riedel, M., Spaemann, R., Wiehl, R., Wieland, W., Editors. *Contemporary German Philosophy*. Volume 2. University Park: The Pennsylvania State University Press, 1982.
de Boer, Theodore. *The Development of Husserl's Thought*. Translated by Theodore Plantinga. The Hague: Martinus Nijhoff, Phaenomenologica, vol. 76, 1978.
Derrida, Jacques. *Edmund Husserl's "Origin of Geometry:" An Introduction*. Translated by John P. Leavey, Jr. Stony Brook: Nicholas Hays, Ltd., 1978.
———. *Speech and Phenomena and Other Essays on Husserl's Theory of Signs*. Translated by David B. Allison. Evanston: Northwestern University Press, 1973.
Dilthey, Wilhelm. *Selected Writings*. Edited and Translated by H. P. Rickman. Cambridge: Cambridge University Press, 1976.
———. *Pattern and Meaning in History*. Edited by H. P. Rickman. New York: Harper & Brothers, 1962.

Bibliography

Drake, Stillman. *Galileo Studies; Personality, Tradition, and Revolution.* Ann Arbor: The University of Michigan Press, 1970.

Dreyfus, Hubert L., Editor. *Husserl, Intentionality, and Cognitive Science.* Cambridge: The MIT Press, 1984.

Elliston, Frederick, A., and McCormick, Peter, Editors. *Husserl: Expositions and Appraisals.* Notre Dame: University of Notre Dame Press, 1977.

Elveton, R. O., Editor and Translator. *The Phenomenology of Husserl: Selected Critical Writings.* Chicago: Quadrangle Brooks, 1970.

Fuchs, Wolfgang Walter. *Phenomenology and the Metaphysics of Presence.* The Hague: Martinus Nijhoff, Phaenomenologica, vol. 69, 1976.

Gadamer, Hans-Georg. *Philosophical Hermeneutics.* Edited and Translated by David E. Linge. Berkeley: The University of California Press, 1976.

Gurwitsch, Aron. *Phenomenology and the Theory of Science.* Edited by Lester Embree. Evanston: Northwestern University Press, 1974.

———. *Studies in Phenomenology and Psychology.* Evanston: Northwestern University Press, 1966.

Husserl, Edmund. *A. Voigt's "Elemental Logic," in Relation to My Statements on the Logic of the Logical Calculus.* Translated by Dallas Willard. *The Personalist* 60 (1979): 26—35.

———. *Cartesian Meditations: An Introduction to Phenomenology.* Translated by Dorion Cairns. The Hague: Martinus Nijhoff, 1973.

———. *The Crisis of European Sciences and Transcendental Phenomenology.* Translated by David Carr. Evanston: Northwestern University Press, 1970.

———. *The Deductive Calculus and the Logic of Contents.* Translated by Dallas Willard. *The Personalist* 60 (1979): 7—25.

———. *Experience and Judgment: Investigations in a Genealogy of Logic.* Edited by Ludwig Landgrebe. Translated by Karl Ameriks and James S. Churchill. Evanston: Northwestern University Press, 1973.

———. *Formal and Transcendental Logic.* Translated by Dorion Cairns. The Hague: Martinus Nijhoff, 1969.

———. *Husserliana.* Band XIX/1 and XIX/2. *Logische Untersuchungen.* Edited by Ursula Panzer. The Hague: Martinus Nijhoff Publishers, 1984.

———. *Husserliana.* Band XII. *Philosophie der Arithmetik.* Edited by Lothar Eley. The Hague: Martinus Nijhoff Publishers, 1970.

———. *Husserliana.* Band XXI. *Studien zur Arithmetik und Geometrie. Texte aus dem Nachlass.* Edited by Ingeborg Strohmeyer. The Hague: Martinus Nijhoff Publishers, 1983.

———. *Ideas Pertaining to a Pure Phenomenology and to a Phenomenological Philosophy.* First Book. Translated by Fred Kersten. The Hague: Martinus Nijhoff, 1983.

———. *Ideas Pertaining to a Pure Phenomenology and to a Phenomenological Philosophy.* Second Book. Translated by Erazim V. Kohák et al., Seminar on Husserl's Second Ideas, Boston University Department of Philosophy, Fall 1978. Unpublished.

———. *Ideas Pertaining to a Pure Phenomenology and to a Phenomenological Philosophy.* Third Book. Translated by Ted E. Klein and William E. Pohl. The Hague: Martinus Nijhoff, 1980.

———. *The Idea of Phenomenology.* Translated by William P. Alston and George Nakhnikian. The Hague: Martinus Nijhoff, 1973.

———. *Introduction to the Logical Investigations.* Edited by Eugen Fink and Translated by Philip J. Bossert and Curtis H. Peters. The Hague: Martinus Nijhoff, 1975.

———. *Logical Investigations,* 2 vols. Translated by J. N. Findlay. London: Routledge & Kegan Paul Ltd., 1970.

——. *The Paris Lectures*. Translated by Peter Koestenbaum. The Hague: Martinus Nijhoff, 1970.
——. *Phenomenological Psychology*. Translated by J. Scanlon. The Hague: Martinus Nijhoff, 1977.
——. *The Phenomenology of Internal Time Consciousness*. Edited by Martin Heidegger. Translated by James S. Churchill. Bloomington: Indiana University Press, 1964.
——. *Philosophie als strenge Wissenschaft*. Frankfurt a.M.: Vittorio Klostermann, 1981.
——. *Psychological Studies in the Elements of Logic*. Translated by Dallas Willard. *The Personalist* 58 (1977): 295—320.
——. Review of Ernst Schröder's "*Vorlesungen über die Algebra der Logik*." Translated by Dallas Willard. *The Personalist* 59 (1978): 115—43.
——. *Zur Phänomenologie der Intersubjektivität. Texte aus dem Nachlass*, Husserliana XIII, XIV and XV. Edited by Iso Kern. The Hague: Martinus Nijhoff, 1973.
Jungnickel, Christa, and McCormmach, Russell. *Intellectual Mastery of Nature*, vol. 2: *The Now Mighty Theoretical Physics, 1870—1925*. Chicago: The University of Chicago Press, 1986.
Kline, Morris. *Mathematics and the Search for Knowledge*. New York: Oxford University Press, 1985.
——. *Mathematics and the Physical World*. New York: Thomas Y. Crowell Co., 1959.
——. *Mathematics: The Loss of Certainty*. New York: Oxford University Press, 1982.
Kisiel, Theodore J. and Kockelmans, Joseph J., Editors. *Phenomenology and the Natural Sciences*. Evanston: Northwestern University Press, 1970.
Kneale, Martha and Kneale, William. *The Development of Logic*. Oxford: The Clarendon Press, 1962.
Knight, David. *The Age of Science*. Oxford: Basil Blackwell, Ltd., 1986.
Kohák, Erazim. *Idea and Experience: Edmund Husserl's Project of Phenomenology in Ideas I*. Chicago: The University of Chicago Press, 1978.
Landgrebe, Ludwig. *The Phenomenology of Edmund Husserl: Six Essays by Ludwig Landgrebe*. Edited and Translated by Donn Welton. Ithaca & London: Cornell University Press, 1981.
Levinas, Emmanuel. *The Theory of Intuition in Husserl's Phenomenology*. Translated by André Orianne. Evanston: Northwestern University Press, 1973.
Levin, David Michael. *Reason and Evidence in Husserl's Phenomenology*. Evanston: Northwestern University Press, 1970.
Lindenfeld, David F. *The Transformation of Positivism: Alexius Meinong and European Thought, 1880—1920*. Berkeley: University of California Press, 1980.
Merleau-Ponty, Maurice. *The Primacy of Perception*. James M. Edie, Editor. Northwestern University Press, 1964.
Mohanty, J. N. *Husserl and Frege*. Bloomington: Indiana University Press, 1982.
——. *Edmund Husserl's Theory of Meaning*. The Hague: Martinus Nijhoff, Phaenomenologica, vol. 14, 1976.
——. ed. *Readings on Edmund Husserl's "Logical Investigations."* The Hague: Martinus Nijhoff Publishers, 1977.
Murphy, Richard T. *Hume and Husserl*. The Hague: Martinus Nijhoff, Phaenomenologica vol. 79, 1980.
Miller, J. Philip. *Numbers in Presence and Absence: A Study of Husserl's Philosophy of Mathematics*. The Hague: Martinus Nijhoff Publishers, Phaenomenologica, vol. 90, 1982.
Nabert, Jean. *Elements for an Ethic*. Translated by William J. Petrek. Evanston: Northwestern University Press, 1969.
Natanson, Maurice, Editor. *Phenomenology and the Social Sciences*, 2 vols. Evanston: Northwestern University Press, 1973.

Ricoeur, Paul. *Freedom and Nature: The Voluntary and the Involuntary.* Translated by Erazim V. Kohák. Northwestern University Press, 1966.

——. *Husserl: An Analysis of his Phenomenology.* Translated by Edward G. Ballard and Lester E. Embree. Evanston: Northwestern University Press, 1967.

Sallis, John, Editor. *Husserl and Contemporary Thought.* Atlantic Highlands: Humanities Press, 1983.

——. *Studies in Phenomenology and the Human Sciences.* Atlantic Highlands: Humanities Press, 1979.

Scheler, Max. *Formalism in Ethics and Non-Formal Ethics of Values.* Translated by Manfred S. Frings and Roger L. Funk. Evanston: Northwestern University Press, 1973.

——. *The Nature of Sympathy.* Translated by Peter Heath. Hamden: Archon Press, 1970.

Schutz, Alfred. *The Phenomenology of the Social World.* Translated by Frederick Lehnert and George Walsh. Northwestern University Press, 1967.

Sokolowski, Robert. *The Formation of Husserl's Concept of Constitution.* The Hague: Martinus Nijhoff, Phaenomenologica, vol 8, 1970.

——. *Husserlian Meditations.* Evanston: Northwestern University Press, 1974.

——. *Presence and Absence: A Philosophical Investigation of Language and Being.* Bloomington & London: Indiana University Press, 1978.

Spiegelberg, H. *The Phenomenological Movement.* The Hague: Martinus Nijhoff, Phaenomenologica, vol. 5/6, 1971.

Stein, Edith. *On the Problem of Empathy.* The Hague: Martinus Nijhoff, 1970.

Theunissen, Michael. *The Other.* Translated by Christopher Macann. Cambridge & London: The MIT Press, 1984.

Tragesser, Robert S. *Husserl and Realism in Logic and Mathematics.* Cambridge: Cambridge University Press, 1984.

——. *Phenomenology and Logic.* Ithaca and London: Cornell University Press, 1977.

Waismann, Friedrich. *Introduction to Mathematical Thinking.* New York: Harper & Row, Publishers, 1951.

Welton, Donn. *The Origins of Meaning: A Critical Study of the Thresholds of Husserlian Phenomenology.* The Hague: Martinus Nijhoff Publishers, Phaenomenologica, vol. 88, 1983.

Weyl, Hermann. *Philosophy of Mathematics and Natural Science.* Princeton: Princeton University Press, 1949.

Whitehead, Alfred North. *An Introduction to Mathematics.* New York: Oxford University Press, 1972.

Willard, Dallas. *Logic and the Objectivity of Knowledge.* Athens: Ohio University Press, 1984.

Wood, Robert E. *Martin Buber's Ontology.* Evanston: Northwestern University Press, 1969.

Index

The entries below are salient topics considered in the text; minor themes mentioned only in passing are not included. Similarly, the names appearing are of those with whose works dialogue is pursued at some length.

Abstracta, see Concepts
Abstraction, of concept of collective connection, 31f.
   Husserl's doctrine of, 32—3
   Husserl's rejection of, 91—102
   of number concepts, 38—40
   role in intuition, 74—7
   role with respect to Ideas, 91—102
Attention, psychical function of, *see* Abstraction

Biemel, Walter, 134—5
Brentano, Franz, 3—5, 7
   Husserl's critique of, 113—15

Causality, epistemological, *see* Explanation
   causal world as "interpretation" by consciousness, 121—4
Certainty, epistemic, 4—8
Collective Connection, characterization of, 24
   concept of, 31f.
Concepts, as proto-Ideas, 91—5
   structure, 50—4
Contents, content relations, *see* Totalizing act
   Inhalten versus *Dingen*, 77—82
   of psychical acts, *see* Totalizing act

de Boer, Theodore, 136—8

Empiricism, Husserl's rejection of theory of sense object, 111—15
   Husserl's rejection of theory of abstraction, 99—102
Explanation, as epistemological correlate of "description," 2—4
   explanatory constructs, *see* Causality

Figural moments, *see* Quasi-qualities
Frege, Gottlob, 141

Gestalt, *see* Quasi-qualities

Ideas, apprehension, *see* Abstraction; as transfigured concepts, 91—5
Immanence, Husserl's rejection of immanent object, 111—15
   immanent object *qua* intuited, 74—86
   psychical, 1—6
   role of "inner perception," 118—21
Intentionality, *see* Representing
Intuition, characterization of, 75—7
   role in totalizing sense object, 72—87

Logic, concepts and laws *qua* Ideatic, 102—6

Miller, J. P., 133, 138—9
Multitude, sensible, characterization, 62—3
   as paradigm for totalization of sensible individual, 74—7
   totalization of, 63—70

Index

Number, concepts, abstraction of, 38—40
  actual (*eigentlich*) and inactual, 54—9
  "attachment" of, 36f.
  generation of, 34—6, 54—9
  qua Ideatic, 96—102
  rationale for translation of (*un*)*eigentlich*, 142—44

Psychical relations, *see* Totalizing act

Quasi-qualities, 67—70

Representation, as psychical extension of totalizing act, 47
  relation to intuitional totalization of sensible individual, 83—6
  role in "production" of transcendent sense object, 115—18
  uneigentlich number concepts, *see* Number concepts
  with respect to totalization of sensible multitudes, 63—70

Sensations, as ensouled or animated, 109—11
Sense object, as modified sensible multitude, 62—3, 74—7
  Husserl's rejection of *qua* immanent, 111—15
  intuition of, 74—86
  transcendent *qua* interpretation, 115—24
Sokolowski, Robert, 139—41
Stumpf, Carl, 5—6
Symbolizing, *see* Representing

Totality, as tantamount to totalizing act, *see* Totalizing act
  rationale for translation of *Inbegriff* as, 129—30
Totalizing act, as appropriator of Ideas, 90—1, 95—106
  as appropriator of sensible individuals, 62—3, 74—86, 115—24
  as appropriator of sensible multitudes, 63—70
  as psychological precondition for every whole, 26
  as tantamount to the totality, 22—7, 48—50
  characterization, 19—22
  misinterpretation as contentual, 40—3
  rationale for use of term, 129—31
Transcendence, *see* Sense Object

Weierstrass, Karl, 6—7
Willard, Dallas, 125, 132, 140